Microwave Materials and Fabrication Techniques

Artech House

The Artech House Microwave Library

Microwave Materials and Fabrication Techniques

Thomas S. Laverghetta

Copyright © 1984 Second Printing, April, 1985.

ARTECH HOUSE, INC.
610 Washington Street
Dedham, MA 02026

International Standard Book Number: 0-89006-143-2
Library of Congress Catalog Card Number: 84-071819

To Andy.
God's greatest gift.

CONTENTS

ACKNOWLEDGMENTS

Thank-you seems like the smallest phrase in the English language when we are trying to adequately acknowledge the help other people have given us, but many times it is the only way we have of expressing gratitude. I sincerely thank everyone who helped make this book a reality.

There are, as always, those who work above and beyond what they are asked to do. These people deserve the highest degree of thanks. A very special thank you to Cheryl Womack and Carol Hilkey, who retrieved many articles from the library; to Willie Doll, who answered an endless list of questions and supplied an uncountable number of books and articles; to Joanne Zelle, who once again did a beautiful job on the drawings; and to my wife, Pat, who put in many uncomplaining hours typing the manuscript while carrying our baby. Also, I would like to thank Joe Krastel of Anaren Microwave and Chuck Hofius and Harlan Howe of M/A-COM for their valuable information and pictures.

Finally, I would like to thank my family for being patient and loving throughout this whole project. Those four young people and my wonderful wife truly make life worth living.

PREFACE

Most books written on microwaves are concerned with circuit designs. This is, of course, a necessity, since it is imperative that the operation and design of microwave circuitry be known and understood. Unfortunately, too many people are not aware of the vast number and variety of materials available for use in these circuits. Also, the average engineer has no idea what happens to a circuit once he or she turns in a drawing or a piece of artwork for fabrication. The intention of this book, then, is to educate microwave circuit designers in materials and fabrication techniques that will help their microwave designs perform as well on the bench as they do on paper.

Chapter 1 is an introduction that asks four questions: Which laminate should you use, how should you attach components to your circuit, how will the circuit be attached to the ground plane, and how will the circuit be packaged? Each of these questions is answered in general terms with specifics being referred to in later chapters.

Chapter 2 deals with microwave laminates and substrates. Terms used to describe the microwave materials are presented as well as discussions on Teflon® fiberglass laminates, high-dielectric laminates, and alumina, sapphire, quartz, and beryllia substrates.

Chapter 3 discusses metals that are used in microwaves, such as aluminum, copper, silver, gold, indium, tin, and lead. Familiarization with particular metals will give the designer better insight into the packaging techniques to use.

Chapter 4 covers microwave artwork from drawing to layout to films used for both positive and negative artwork. A two-way power divider is used as an example for these discussions and for future discussions in later chapters.

Chapter 5 takes the artwork generated in Chapter 4 and performs the etching process. Topics covered are photo resists, artwork placement and exposure, and etching methods. Once again, both positive and negative methods are presented as well as procedures for both laminates and substrates.

Chapter 6 contains some of the most important information needed to make a design perform properly. This chapter is on bonding techniques and covers solders, epoxy, and bonding methods (thermocompression, ultrasonic, and thermosonic). Component and substrate attachment are also presented.

Chapter 7 and 8 are concerned with packaging and connectors and transitions, respectively. Microstrip and stripline packaging techniques are presented as well as a variety of connectors and methods of making the transition from these connectors to the circuit.

As stated previously, the intention of this book is to educate the microwave circuit designer in materials and fabrication techniques. If a designer is familiar with the techniques discussed in each chapter in this book, he or she will certainly avoid many problems and will have a design that will perform on the bench as it does on paper — maybe even better.

Thomas S. Laverghetta
Auburn, Indiana

April 1984

CHAPTER 1
INTRODUCTION

A microwave circuit design, when completed, can be looked upon with great pride. It has required many hours of analysis, evaluation, and optimization to arrive at that final product — your own special microwave circuit. There are, however, many questions to be answered once you have completed this design and are able to prove its worth: What laminate should you use? How will you attach components to the circuit? How will the circuit be attached to the ground plane? How will the circuit be packaged? (You will note that all these questions refer to stripline or microstrip construction. This will be true throughout this text. Whenever a procedure refers to coaxial or waveguide construction it will be called out as such. Otherwise, all processes and procedures will be for stripline or microstrip.)

The questions asked above may seem trivial at first glance because you may think that the most difficult task, that of design, is over. Actually, this difficult task is still continuing because the design of a microwave circuit is not complete until it has been given an appropriate case.

To show just how important this last statement is, we will once again pose the above questions and answer each with one or more examples. Consider the first question: What laminate should you use? The microwave laminate, which will be covered in detail in Chapter 2, is the printed circuit board of

1

microwaves. Although we refer to it as such, it is much more than simply a support for printed lines, as is the case with a PC board. Its parameters are critical to circuit operation. Such parameters as dielectric constant, dissipation factor, and dielectric thickness all must be held within tenths or hundredths. (All of these terms will be covered in detail in Chapter 2.) Also, the choice of the laminate according to dielectric constant is an important factor. Consider the following example. Suppose you have an area 1/2 in. by 1/2 in. in which to put a 50Ω line, which is to be one wavelength at 2 GHz. This area will be in stripline. Suppose you had two laminates available to you. One had a dielectric constant of 2.1, and the other, 10.2. The first step in determining the proper laminate would be to find out how long a wavelength is for each of the dielectric constants. Using the common wavelength formula below, we calculate these values:

$$\lambda = \frac{c}{f \sqrt{\epsilon}}$$

$$\lambda_{\epsilon=2.1} = 4.07'' \quad \lambda_{\epsilon=10.2} = 1.85''$$

It is obvious from these numbers that we will get 1.85 in. of line on our laminate much easier than 4.07 in. To show that this is possible, refer to Figures 1.1 and 1.2.

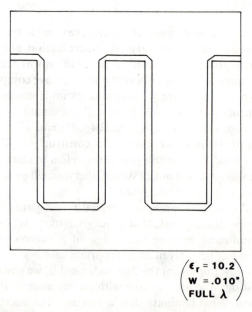

$$\left(\begin{array}{l} \epsilon_r = 10.2 \\ W = .010'' \\ FULL\ \lambda \end{array}\right)$$

Figure 1.1 Full wavelength on high-dielectric laminate.

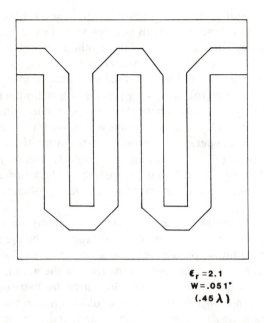

ϵ_r =2.1
W =.051"
(.45 λ)

Figure 1.2 Fraction of a wavelength on high-dielectric laminate.

Figure 1.1 is a layout of the complete wavelength at 2 GHz using a laminate with a dielectric constant (ϵ) of 10.2. You can see that it fits easily on the substrate. On the other hand, Figure 1.2 shows the line using a material with a dielectric constant of 2.1 and obviously will not make it, since 0.45 λ takes up the entire substrate. It is necessary to consider your particular circuit and how it will be laid out before choosing the microwave laminate that will be used.

The second and third questions will be combined, since they are very closely related. The second question concerns attaching components to the circuit, and the third, attaching the circuit to the ground plane. Any time you attach two surfaces together you must be careful as to which metals are being joined. Also, you must be aware of the metallic content of the substance being used to join the surfaces. Most microwave circuits or components are attached using either solder or conductive epoxy. Both materials have specific metallic contents that must be checked for possible problems with the surfaces they are bonding.

Solders can contain a variety of metals in a variety of combinations. The most common is tin/lead. A quick glance at available tin/lead (SnPb) solder shows approximately 19 different combinations are available. The temperature ranges for these solders range from 361°F (182.7°C) to 608°F (320°C).

This range is a result of the percentage of both tin and lead in the solder. By varying the ratio of these two metals you can vary the melting temperature. The solder used will depend on particular application and the temperatures that can be withstood by the components to be attached.

Elements other than tin and lead are used in solders for microwave circuits. Such metals as indium (In), silver (Ag), gold (Au), cadmium (Cd), and zinc (Zn) are commonly used for a variety of reasons, depending on each individual metallic interface. (That is, if you have a component or substrate with a certain metal on it (copper, gold, etc.) and a base plate with a certain metallic contact (aluminum, gold, or tin plating, etc.), the interface between these two surfaces must be compatible. This interface is the solder you are using; it must be an efficient interface and not introduce any new problems because of its construction.)

An example of a problem that can result from an improper solder occurs when a tin/lead solder is used on gold. If, for example, you use 60/40 tin/lead solder (60 percent tin, 40 percent lead) on a gold metallized substrate, there will be a metallic interaction between the tin and the gold that results in a brittle joint that will not be very reliable. There are two solutions to this problem. First, and most obvious, use a solder with no tin in it. This, of course, eliminates any metallic interaction and, thus, any problem. Many times indium solders are used as alternatives: for example, a solder of 50 In 50 Pb (50 percent indium, 50 percent lead). There are a variety of other types available that do the job without having any tin content, but tinless solder is not always an available alternative. If it is not, use the second solution, which is a solder with a small content of silver. (A typical ratio is 62.5 percent tin, 36.1 percent lead, and 1.4 percent silver.) The silver will inhibit the interation, or leaching, of the gold and tin. Since there usually is more than one solution to any problem you may encounter, it is important to check the metallic components of substrates and component terminations to be sure there will not be any intermetallic interface problems. It is a good idea to consult the component data sheet and heed the advice of the vendor as to the recommended method for attaching the component.

As previously mentioned, the substrates and components can be attached with solder or epoxy. Epoxies, which will be covered in detail in Chapter 6, are used quite frequently for attaching small chip components to a substrate or transistors in chip form to a carrier. It is a precise method for attaching these tiny parts without having to subject them to heat as high as some solders would require. Most curing temperatures and times are in the neighborhood of 80°C (170°F) for one and one-half hours; 120°C (248°F) for one hour; or 150°C (302°F) for one-half hour for two-part (epoxy resin and hardener) epoxy. For one-part epoxy temperatures and times range from 150°C (302°F) for two hours to 170°C (338°F) for one hour.

When using epoxies be sure to clean each of the surfaces to be joined. This eliminates any contaminates that the epoxy will have a tendency to cling to. These contaminates may form a good bond to the epoxy, but the electrical conductivity will be very spotty. Also, do not place epoxy on bare aluminum or to a pure gold surface. The nature of epoxy is to rely for adhesions on chemical groups of atoms, termed *radicals*, that contain hydrogen and oxygen. Neither pure gold or the bare aluminum has these hydroxyl radicals, and thus neither results in good adhesions. An iridite, or alodine, treatment or the appropriate plating will supply the necessary radicals for proper adhesion.

The fourth, and final, question that was asked was, How will the circuit be packaged? Breadboards or first units often are built on an aluminum plate to check out the design. The next step is to put them into a final package configuration so that finished units may be fabricated. This simple step sometimes turns into an ugly nightmare if careful thought is not put into the package design.

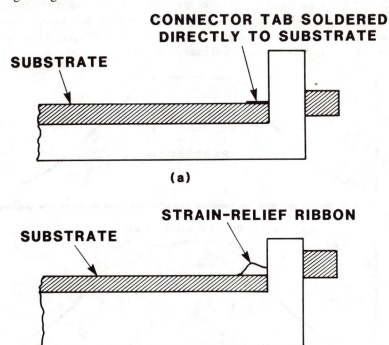

Figure 1.3 Substrate-to-connector transitions.

Consider two areas of crucial importance when designing a package: substrates-to-connector (microstrip-to-coaxial) transitions and package height. To illustrate the importance of the proper substrate-to-connector transition, refer to Figure 1.3. (This example is for microstrip circuits, but similar precautions are needed for stripline circuits. All cases will be covered in Chapter 8 of this text.) Figure 1.3a shows a side view of a microstrip substrate in an aluminum case with the connector tab soldered directly to the circuit. This particular circuit was a bandpass filter and was run over its required band with the connectors soldered directly to the substrate. Its response is shown in Figure 1.4a. As can be seen, it follows a good band shape with good roll off on both the lower and upper sides of the passband. At this point it was feared that the extreme temperature range that the filter would be subjected to (–40°C to +60°C) might cause the solder joint to fracture unless a mechanical stress relief was used. Thus, the arrangement shown in Figure 1.3b was used with a ribbon put between the connector and the substrate. This resulted in an excellent stress relief mechanically but caused the problem shown in Figure 1.4b. The high end of the filter response developed a definite hump, which was very undesirable. This was traced back to the ribbon, which added a significant amount of inductance to the transition.

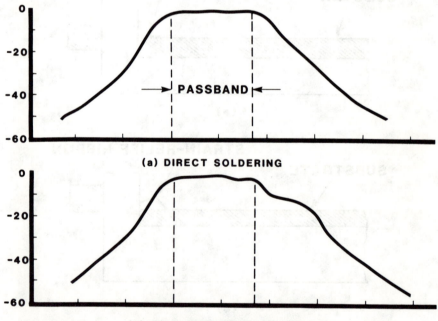

(a) DIRECT SOLDERING

(b) WITH STRESS-RELIEF RIBBON

Figure 1.4 Bandpass filter response with different transition methods.

The above example illustrates that a fix that will solve mechanical problems will not necessarily keep electrical problems from occurring at the same time. Do not take the connector transitions for granted. Design them as carefully as you do the circuit to which they are connected.

The second area to be concerned with in circuit packaging is the height of the case. As mentioned before, many times you will make your first breadboard units on a flat plate with no sides or cover. The design is finalized in this configuration and then put into a box. Suddenly your response shifts, cuts off on the high end, or develops a large resonance in the center of your band of operations. You have forgotten, or were not aware, that the cover and sides of a case can set up definite, and different, modes that can change operations drastically.

To overcome this you should keep the cover at least 10 ground-plane spacings above the circuit. That is, if your material is 0.030 in., the cover should be at least 0.300 in. above the circuit. Also, use separations between circuits wherever possible to break up any waveguide effects that can occur. These basic precautions can usually help avoid any case problems with your circuits.

We have mentioned only two of the problem areas to watch for when designing a package for a microwave circuit. There are other effects to consider, but these are representative of the type of problem encountered and are sufficient for an introduction. The other area will be covered in detail in Chapter 8 of this text.

This chapter began by saying that there are many questions to be answered once you have completed your design. Four very important questions were posed and introductory answers provided to each of them. It should be clear by this point that the fabrication of that beautiful circuit you have designed takes at least as much thought and design time as the original circuit did. Remember, that circuit is totally useless sitting on your desk as a circuit board. To prove it out you have to use the proper laminate, properly attach components to the laminate, connect the laminate or substrate to the case, efficiently get from the circuit to the connector, and have the proper case design. Fabrication is another complete design exercise.

CHAPTER 2

LAMINATES AND SUBSTRATES

2.1 INTRODUCTION

Microwave laminates and substrates are, in a very broad sense, the printed circuit boards of microwaves. They do much more, however, than provide mechanical support for circuit trances, which is basically what the conventional PC board does. They are actually an integral part of the very circuits they support, since their characteristics determine the length and widths of the traces used for microwave applications. The terms *laminate* and *substrate* are generally used interchangeably. Usually when laminate is used people are referring to the Teflon^R fiberglass materials that have the lower-dielectric constants and copper on them. Conversely, when substrates are referred to, the inference is usually to the hard ceramic materials that have a metallization (gold, chrome, copper, etc.) on their surfaces. As we have said, though, the terms are interchangeable, and you should be sure you know which an individual is referring to when discussing these materials. Throughout this text we will usually present both terms.

Important parameters for microwave laminates and substrates are dielectric constant, dielectric thickness, and dissipation factor. We can see how important the first two parameters are by referring to the expression below for the impedance of strip transmission line,

$$Z_o \sqrt{\epsilon} = 60 \log_e \left(\frac{4b}{\pi d} \right)$$

where

Z_o = characteristic impedance of the line

ϵ = dielectric constant of the laminate or substrate (a detailed definition follows later)

b = thickness of the ground-plane spacing (twice the thickness of the laminate for stripline; the dielectric thickness for microstrip)

d = a relationship that accounts for line width, w, and copper thickness, t.

From this expression you can see how important the dielectric constant (ϵ) and thickness are to the impedance of a microwave transmission.

The values of the basic parameters mentioned above (dielectric constant and thickness) are important to a design. The tolerances on these parameters are also important. There are many times when the tolerances are actually of more importance than the actual value of the parameters. This should not be too hard to understand if you refer to the impedance expression we used before. You can see how the impedance would vary if the dielectric constant or material thickness fluctuated to any extent. Consider, for example, the following specifications for a typical piece of G-10 glass epoxy PC board material:

Dielectric constant   4.0 to 4.6

Thickness   0.031 in. $\pm$ 0.004

                          0.062 in. $\pm$ 0.006

If we used these fluctuations in parameters for our calculations we would find some drastic variations. To illustrate this we will use them in the impedance formula shown earlier to see what happens to only one of the circuit parameters, Z. Conditions to be used are:

ϵ = 2.55

b = 0.031 in

w = 0.046 in (width of line)

t = 0.0014 in (thickness of copper)

Z_o = 50

By substituting different dielectric constant and thickness values into the impedance formula, a range of values can be achieved. The results of this substitution are shown in Table 2.1.

From this table you can see that any variation in dielectric constant (ϵ) causes a corresponding variation in line impedance. It can easily be understood, therefore, that the closer you hold the dielectric to that used in the initial calculations, the closer you stay to the desired impedance. Also, it is interesting to note that the percentage variation in impedance is greater than

Table 2.1
Material Comparison

Specification	$\Delta\epsilon$	Δz	Δb	Δz
G-10	± 7.5%	**3.77 Ω** (7.5%)	± 0.004 in.	9.77 Ω (19.5%)
Microwave laminate	± 2%	1.0 Ω (2%)	± 0.001 in.	2.5 Ω (5%)
Laminate				

ϵ = Dielectric constant
z = Impedance
b = Dielectric thickness

the percentage variation in dielectric thickness. A 13 percent variation in thickness (± 0.004 in), for example, causes a 19.5 percent change in impedance. Similarly, a 3 percent thickness variation (± 0.001 in) will result in a 5 percent change in impedance. You should be able to see very clearly from this that it is of the utmost importance that the dielectric thickness be held as constant as possible. Actually, whenever working with microwave circuits you should have laminates or substrates that have all parameters held to very close tolerances. This will ensure that when an impedance or length of microstrip or strip line is calculated, it is fairly certain to come out to be that value when it is built and tested.

We have used terms in these first few paragraphs that describe microwave laminates and substrates. We are now ready for detailed explanations of these terms so that you can become even more familiar with the microwave materials that are such an integral part of virtually all microwave circuits. Terms to be covered are *dielectric constant, dissipation factor, dielectric thickness*, and material *cladding*.

To understand the term *dielectric constant* we must first define the term *dielectric*. The formal dictionary definition of dielectric is "an insulation or nonconducting medium." This definition immediately brings to mind something that blocks or obstructs energy or keeps it from traveling through a material. This, in fact, tells much of the story of a microwave dielectric. To aid in understanding dielectrics, consider the following example. Suppose you drove your car to and from work along the same route every day. Along this route you drove through a low spot in the road. On clear dry days you drive the entire route virtually at the same speed and unobstructed. If on some days it rains and the low spot collects three to four inches of water, you will be slowed down as you drive through the water. This will obstruct your progress compared to when the road was clear. Also, during the winter it may snow five or six inches, and your speed will be reduced even further. Once again your progress and speed will be obstructed.

In each of the above cases the car was slowed down by means of obstruction. The car did not change; only the environment the car encountered changed. In the first case the environment is simply air around the tires of the car; in the second case it was water around the tires; and the final case it was snow causing a great resistance (or obstruction) to the tires. In the same manner a microwave laminate causes the microwave energy to be subjected to a different environment that slows down, or obstructs, the energy to change such parameters as velocity, wavelength, and delay. It is this environment, which is different from air, that we call a dielectric:

A *dielectric* is a material that creates an environment causing microwave energy to be reduced in velocity from what it would be in free space.

The term *dielectric constant* has basically been defined in our example. As you will recall, we compared the degree of difficulty in going through the water and snow by referring to the speed we would be going when it was a clear dry day. Similarly, we have compared the velocity of microwave energy in a dielectric to what it would be in air. Also, we make calculations of a free-space wavelength (air), and the wavelength in a dielectric medium. In each case there are two numbers: One is for air, and one is for the dielectric we are using. Each of these numbers is a constant: The value in air is 1.0, and the value through the dielectric is a constant depending on the physical properties of the materials used. (We call it a constant number, but it will vary slightly by the tolerance values previously mentioned.) If we take these two numbers and form a ratio we will come up with the term *dielectric constant*. Very simply put, it is the ratio of the conditions encountered in a dielectric environment to those encountered in a pure air environment. The symbol for this relative value is ϵ_r (ϵ_{air} = 1.0).

When we hear the word *dissipation* we usually think of something lost. When you think of heat dissipation, for example, you can visualize heat pouring from a radiator or furnace. When you hear on the news that the early morning fog will dissipate by 10 A.M., you picture it as gone by that time. The term *dissipation* in microwaves has a similar meaning. As we have previously said, dissipation is a loss. In microwaves, dissipation is the loss of energy in the form of heat. Thus, in the case of microwave laminates and substrates, we want the material to store energy within the structure rather than dissipate it. This storing of energy will result in a low loss path for the microwave energy. (When we speak of storing energy in the laminate or substrate we mean that the energy remains in the circuit rather than being sent off as heat and causing the material to exhibit a high loss.)

The term *dissipation* is not generally the term used when microwave laminates or substrates are being characterized. *Dissipation factor* is the more recognized phrase and appears on all microwave material data sheets; it defines the loss characteristics of the material. This phrase (termed Tan δ) is actually a ratio. It is the ratio of energy dissipated to energy stored:

$$\tan \delta = \frac{Energy\ Dissipated}{Energy\ Stored}$$

(Tan δ is also termed the tangent of the loss angle of the material. The loss angle is determined by the relationship between loss term and the conductivity term of a material).

If we consider the energy dissipated as a loss, or resistance, and energy stored to be the measure of conductivity of the material, we have a relationship where:

$$Dissipation\ Factor = \frac{Loss\ (Resistance)}{Conductivity}$$

It should be obvious from our discussions that to have an efficient laminate or substrate you need a low value of dissipation factor. From the relationship above you can see that this is possible when the resistance loss is low and/or the conductivity is high. Some representative dissipation factors are shown below:

Material	Dissipation Factor (Tan δ)
Glass epoxy	0.0180
PTFE* glass	0.0018
Alumina	<0.0001
Polyethylene	0.0002
Mica	0.0026
Silicon rubber	0.0032
Ethyl alcohol	0.0620

*PYFE = Polytetra fluorthylene.

As previously stressed, the dielectric thickness of a microwave laminate or substrate is of prime importance. The examples shown at the beginning of the chapter illustrated how microwave parameters will vary greatly with variations in dielectric thickness. To really appreciate the importance of the dielectric thickness you must be aware of exactly what thickness we are referring to. When you look at a microwave laminate or substrate you will see

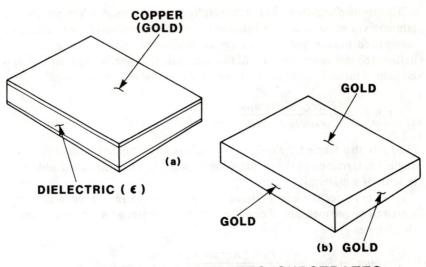

MICROWAVE LAMINATES/SUBSTRATES

Figure 2.1 Microwave laminates and substrates.

DIELECTRIC THICKNESS

Figure 2.2 Dielectric thickness dimension.

a piece of material with copper on both sides and some dark brown or white material in the center; gold completely surrounding a base material; or a white base material with gold on two sides. These are shown in Figure 2.1. Figure 2.1(a) shows the laminate or substrate with copper (or gold) on both sides but not on the edges. Figure 2.1(b) is a representation of a substrate with gold metallization on all six surfaces. (Actually, there is more than just gold on the substrate, but this will be discussed in detail later in this chapter.) From these drawings you could arrive at a thickness, but it would be the wrong thickness. The dimension we want is shown in Figure 2.2. You can see that it is the dimension of the dielectric *only*, not the total laminate or substrate. Any cladding or metallization dimension must be subtracted from the total dimension. Thus, when the data sheet says that laminates are available in 0.031 in and 0.062 in sizes, they are referring to only the dielectric material dimension with no cladding or metallization. This is the *unclad laminate dimension* . Be sure that when checking the dimension of a laminate or substrate you do not confuse the *total* dimension with the *dielectric thickness*. This confusion can cause serious problems with your calculations and thus your final results.

We mentioned *cladding* above when we talked about the dielectric thickness of a microwave laminate or substrate. We will now discuss this term and also expand on it by covering *metallization*. Generally speaking, when the term *cladding* is used it refers to microwave laminate with a copper cladding on one or both sides. *Metallization* is used when we refer to the covering on ceramic (alumina), sapphire, or quartz substrates. We will discuss laminate cladding first.

When you talk about cladding you must give two very important pieces of information with the term: the weight of the metal (copper) and how it is fabricated for the laminate (electrodeposited or rolled). The weight of the copper as specified on a laminate data sheet usually has standard values of 1/2 oz, 1 oz, and 2 oz. These numbers refer to the fact that one square foot of copper will weight 1/2 oz, 1 oz, or 2 oz, depending on what weight you choose. Each of these weights, logically, has a definite thickness associated with it: Three different weights of copper, all one square foot in size, must be different in their thickness. The relationship between weight and thickness is shown below:

Weight (ounces)	Thickness (inches)
0.5	0.0007
1.0	0.0014
2.0	0.0028

A 1/4-oz copper is also available from most manufacturers. This, however, is usually a special order. The 1/4-oz may be needed for very narrow lines or when narrow gaps between lines are required. Generally, however, the 1/2-oz copper will suffice and is a standard cladding. In Figure 2.3, Case I, we have a laminate with 2-oz copper at the left and either 1/2- or 1/4-oz copper at the right. (Since we have said that generally the 1/2-oz copper is used because it is a standard cladding as opposed to the special-order 1/4-oz, we will make all of our comparisons with 2-oz and 1/2-oz from this point forward.) In this case we are required to etch a very narrow gap (<0.005 in) between two transmission lines. You can see that the lines etched on the 2-oz copper are dramatically undercut — that is, they are narrower on the underside of the line than at the top. This is because of the time required to etch the thicker copper. The etchant does not etch the top portion because there is photo resist on that surface. The lower portion is thus subjected to the etchant for a long duration of time and over etches. The 1/2-oz copper portion is virtually undisturbed, as can be seen in the figure, because of the smaller volume of copper that needs to be etched to arrive at two transmission lines separated by a narrow gap.

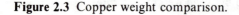

Figure 2.3 Copper weight comparison.

In Case II we have only a single line to etch, which is narrow (<0.010 in). The same conditions occur when 2-oz copper is used, as can be seen in the figure. Once again there is an undercutting that causes variations in dimensions of the line. You can see also that the 1/2-oz copper is virtually undisturbed when used for narrow lines. One condition must be watched in this case and in Case I discussed above: how drastic this undercutting becomes. If the circuit is etched for too long a period of time the undercutting will be so severe that the line lifts completely away from the laminate. This is shown in Figure 2.4. Care should be taken when etching narrow lines, whether on thick or thin copper.

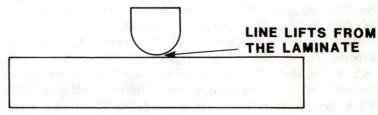

Figure 2.4 Overetching of lines.

As a summary of our copper weights discussions for microwave laminates, consider the following advantages and disadvantages.

2-oz copper
- Good copper weight for high-power applications;
- Copper will undercut on narrow gaps and lines;
- Narrow lines inconsistent along a long length;
- Good for general-purpose circuitry;
- A standard weight.

1-oz copper
- A very common weight for a variety of microstrip and stripline circuits;
- Less undercut on narrow gaps and lines than 2-oz;
- Can be used for medium-power applications;
- A standard weight.

1/2-oz copper
- Used for low-power applications;
- Excellent for narrow gap and line etching;
- Do not use for high-power circuits;
- A standard weight.

1/4-oz copper
- Should be used only for low-power circuits;
- Excellent for narrow gaps and lines;
- A nonstandard weight usually available in special order only.

Now that we have looked at the different weights of copper, let's see how the copper is fabricated. Previously we mentioned two methods of copper fabrication — rolled and electrodeposited. The standard method used for most microwave laminates is electrodeposited (ED) copper. Rolled copper is available by special request, and the data sheet usually has a statement saying the laminate can be supplied with rolled copper "for more critical electrical applications." We will cover the standard ED copper first and then the rolled.

If you were to look up the word *electrodeposition* in the dictionary, either a conventional or electronic dictionary, you would find the typical dictionary definition: It would say that electrodeposition is the process of depositing a substance on an electrode by electrolysis. Unless you are a chemist, this probably tells you no more than you knew when you first looked up the word. To clear the water a little you could look up *electrolysis*. At this point you would find that the dictionary says that electrolysis is the process of changing the chemical composition of a material by sending an electric current through it. If you put these two definitions together and do some thinking, you will get some idea of what ED (electrodeposited) copper is. Here is a basic definition of the term:

> *Electrodeposited copper* is a material produced by a chemical building process in which the individual copper particles are electrically joined to form the desired sheet thickness.

This can be likened to taking small pieces of wet clay and building one continuous sheet from the small pieces. If you had a certain number of these pieces of clay placed next to one another over some base surface (as ED copper is plated on a rotating drum structure and then pulled off), you would have a structure that was a certain thickness and consistency. This might be likened, for example, to plating (or depositing) 1/2-oz copper (0.0007 inches). If you needed 1-oz copper (0.0014 inches) you would repeat the process until the desired 1.4 mils of thickness was achieved. Similarly, you would have to add more wet clay to your already existing structure in order to achieve the required thickness. This depositing process can be very precisely controlled by controlling the time and current used in the electrodeposition process.

As previously mentioned, ED copper is the standard copper found on microwave laminates. Rolled copper is obtained by special request to the manufacturers and is used for more critical applications. You will see why this is so when we complete our discussion on rolled copper and compare it with standard ED copper.

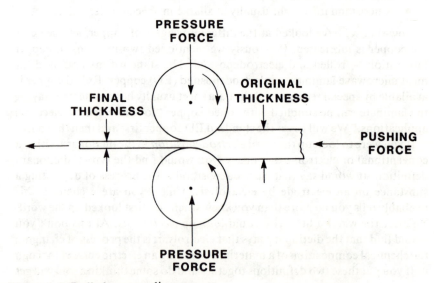

Figure 2.5 Rolled copper diagram.

The easiest way to picture rolled copper is to compare it to putting something through an old-style wringer washer. The copper block that is used to make a copper foil for laminates is compressed in much the same manner. Figure 2.5 shows this concept. The original thickness of copper is pushed into the pair of rollers, which are being both rotated and forced together. (This vertical force depends on the final thickness required at the output.) As the copper is forced through the rollers, it is compressed to this previously determined thickness. Following this compression there are various processes that may be used on the copper to obtain the desired consistency or hardness. For electrical applications it is desired that the copper be relatively soft to increase conductivity in the material. This process, as previously mentioned, is used for more critical applications because the rolled copper is a much more

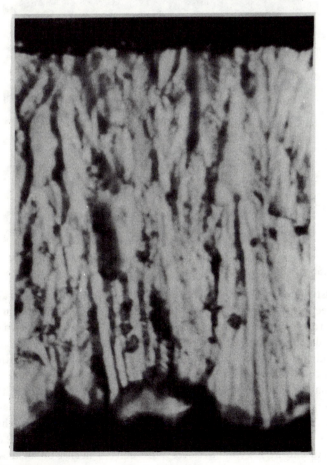

Figure 2.6a ED copper.

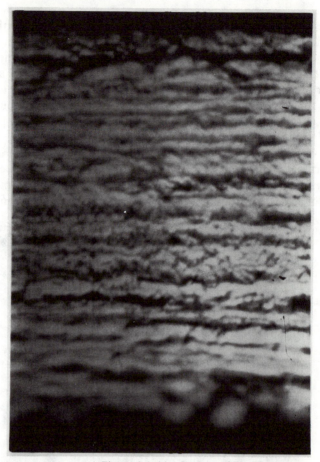

Figures 2.6a&b - Courtesy of The MICA Corp.

Figure 2.6b Rolled copper.

evenly distributed and consistent copper than ED copper. Figure 2.6(a) shows a cross-section of ED copper while Figure 2.6(b) is rolled copper. You can see how the rolled copper is in very even layers, but the ED copper does not exhibit this obvious uniformity. This is not to say that ED copper is not a good conductor at microwave frequencies, but when critical applications are involved it may be wiser for you to use a rolled copper laminate.

Now that we have the weight and fabrication of the copper defined, the next logical step is to put the copper on a microwave laminate. All microwave laminates, whether using ED or rolled copper, are put together, as the name implies, by a lamination process. This basic process is familiar to all of us in one form or another. If you have ever had a driver's license or social security card or business card laminated you will know what we are talking about. Consider the steps involved in laminating one of your cards.

- The card is placed between two plastic laminating sheets;
- A press is heated to the lamination temperature;
- The card and laminating material is placed between the heated plates of the press and pressed together; and
- The card is now laminated and would probably be partially or totally destroyed if you attempted to separate the laminations.

This is the same basic process that is used to put the copper on a microwave laminate. One very important task must be performed before the copper can be laminated to the dielectric material: That is a process of roughing up the surface of the copper so that it will "stick" to the dielectric. During the electrodeposition process a certain amount of roughness is encountered. There usually needs to be additional roughing for good adhesion. Rolled copper usually is very smooth when completed and thus requires more roughing. This surface roughing is usually accomplished by means of an acid etch. This is similar to the roughing necessary when you iridite bare aluminum. The aluminum is subjected to an acid bath first to rough up the surface so that the iridite solution will "stick" to the surface of the aluminum. The same type process is used for copper prior to the lamination process.

When combining copper with a dielectric material the same ingredients are used as with the business card — heat and pressure. With these two ingredients present and controlled, a highly reliable and efficient microwave laminate will result.

We have covered cladding for laminates, now we will cover *metallization* for substrates. The formal definition of metallization is "the deposition of a thin film pattern of conductive material onto a substrate to provide interconnection of components or conductive contact for interconnection." A basic definition of the term is very simply the attachment of a metal to a substrate by means of depositing techniques. When referring to metallization on ceramic substrates (as used in microwave applications) we usually call the process by which we attach the metal to the substrate *sputtering*.

Sputtering is a highly controlled method of coating one material with another (for example, a ceramic substrate coated with a metal). The idea of sputtering is one of energy. The material that is ultimately deposited is literally blasted from a target material by high-energy gas ions. Normally when you think about putting one material on another you think of a direct process in which application is direct, as in painting or a lamination, as we discussed earlier regarding ED and rolled copper. This, however, is not the case with sputtering. The method of application is more of an *indirect* process. To understand this statement, and sputtering, consider the following example.

Supose you had a cardboard box and placed balloons in one end of it as shown in Figure 2.7. The balloons are put in the box side by side so that there is a tight fit. The balloons are considered to be the target. If we now place a hose so that it is in direct line with the center of the balloons and send out a short burst of water, we will simulate the action of sputtering. (The hose and short burst of water represent the gas ion that causes the target material to be placed on the substrate.) The burst of water strikes a balloon (atom) and tries to push it to the back of the box. This cannot be accomplished because of the box itself and the other balloons holding it in place. Instead, the balloon compresses under the pressure of the water burst. When it comes back to its original spherical shape it transmits the impact force to the other balloons around it. The force works its way to the front row where the balloons are no longer restricted but are free to move. They move out in the direction from which the water burst came.

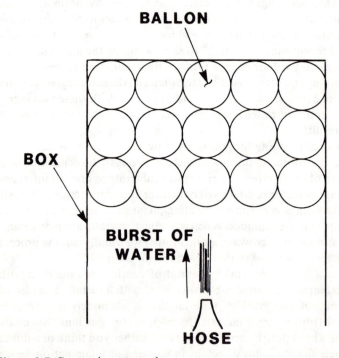

Figure 2.7 Sputtering example.

With the example presented in mind, you can look at Figure 2.8 and see how a basic sputtering process works. The target consists of the material to be deposited on the substrate (gold, for example). Notice that it has a negative (-) charge. The gas ion (usually argon) that bombards the target is shown in the center. The ion hits the target and breaks material loose just as we illustrated with the balloons. This material (atoms) travels through the vacuum to the positively (+) charged substrate. The layer deposited can be very carefully controlled by adjusting the voltage applied to the plasma (ion) and the time of exposure of the substrates.

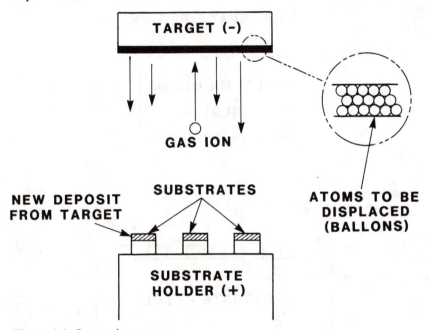

Figure 2.8 Sputtering process.

There are two terms that may come up if you become involved with sputtering or plan to investigate further into its operation: *plasma glow region* and *dark space region*. Figure 2.9 shows where these regions are in the sputtering setup. The *plasma glow region* is produced because the target is negatively charged, which results in electrons being injected into the gas (usually argon) that is around the target. Because of the voltage applied to the system the electrons are accelerated toward the positive charge on the sub- strates and substrate holder. As the electrons travel toward the positive side, they may collide with a gas molecule. When this happens they give up part of their energy and leave behind an ion and an extra free electron. The cumulative effect of this phenomena is a self-sustaining glow discharge. This ionized area actually heats up and glows. This effect can be viewed in a fluorescent tube. The only difference is the gas used (argon for sputtering, neon for fluorescent lights).

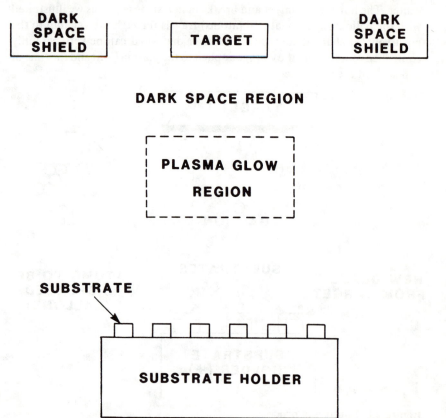

Figure 2.9 Plasma glow and dark space region.

This glow will not occur close to the target, and this is called the *dark space region*. This is because the further the electrons travel, the more chance they have to strike and ionize the argon (gas) molecules. Therefore, very little ionization takes place close to the target, there are no collisions, no plasma glow, and thus a dark area appears.

The *dark space shield* shown in Figure 2.9 takes advantage of the dark space region. This shield (or ring) is placed around the target to ensure that a minimum amount of target material is wasted by being emitted from the side or back of the target. With the shield in place the maximum amount of target material is sent to the substrate where it is intended to be.

Although sputtering has become more prominent in microwave applications recently, it is not a new process by any means. The idea of sputtering was first presented in 1852 by Sir William Robert Grove. At that time he called the process *cathode disintegration*. In 1909 F. Stark gave the first true and accurate description of the process, and in 1921 Sir Joseph John Thompson christened the process *spluttering*. The term *sputtering* resulted when a published paper in 1923 dropped the *L*. In 1925 sputtered films were investigated for use by Western Electric for commercial application. In 1928 Western used the process for manufacture of phonograph records and contacts on microphone transmitters. Until these applications, sputtering was used mostly for decorative effects. These early processes were DC sputtering and were useful only for metallic films. This, of course, is the process used for depositing metal on microwave substrates. In the early 1970s RF sputtering was developed and allowed nonmetals to be sputtered. Advances in vacuum technology have further improved the quality and reproductivity of sputtered materials. Today, the major uses of sputtering are in the electronics industry.

Some of the combinations sputtered for microwave use are shown below. Note that in some cases we say that the materials can either be sputtered or plated. All of these substrates are on alumina material.

Material A
- Chromium (Cr) — adhesive-sputtered
- Copper (CU) — conductor-sputtered
- Gold (Au) — passivation layer plated

Material B
- Chromium (Cr) — adhesive-sputtered
- Gold (AU) — conductor-sputtered/plated

Material C
- Titanium tungsten (TiW) — adhesive-sputtered
- Gold (AU) — conductor-sputtered/plated

You will notice that none of the materials (substrates) listed above is gold directly on the alumina substrate. This is because gold will not stick (or adhere) to the ceramic material. To put gold directly on the ceramic material would be similar to using super glue on wood. The glue soaks into the fibers of the wood, and there is no adhesion. Similarly, there is a migration of the gold into the substrate and no clear boundary between the substrate and the gold. This is a common fault with power transistors where gold metallization is used without the proper buffering and adhesive layer. This migration does not form a bond and a simple piece of a tape will remove the gold from the substrate. For this reason an adhesive layer is needed (chromium or titanium tungsten in the examples shown).

Examples of usage for each of the materials listed are:

Material A: For applications with solder;
Material B: For application where epoxy or bonding techniques are used;
Material C: For high-temperature applications.

Before we get into specific types of laminates and substrates, let us review their history. The history of laminates and substrates for eventual microwave usage began in the 1950s. Early experiments used low-dielectric-constant plastics in a microstrip configuration, but these first attempts were not too successful because the effects of radiation and intercircuit coupling were not taken into consideration.

Polystyrene was found to be a uniform low-loss material but was shown to crack or soften at temperatures above ambient. Attempts to harden and reinforce the material by adding glass fibers destroyed many of the desirable properties it originally had. As an alternative to the polystyrene, PTFE/glass cloth laminates were considered. This material was originally developed for other uses and had some rather unpredictable dielectric properties, but it fulfilled many of the necessary requirements for a microwave laminate.

By the early 1960s PTFE/glass cloth laminates had their dielectric constant reduced from their original 2.65–2.75 to the more popular and usable 2.55 of polystyrene. With this advance the PTFE/glass laminate replaced the styrene product and eliminated many problems. By reducing the glass content even further the dielectric constant was reduced to 2.45 and the first industry-standard laminates were created, becoming the GX type product in MIL-P-13949. (This specification will be presented and explained later in this chapter.)

Dielectric constants were further reduced by the introduction of irradiated polyolefin (2.32) in the 1960s. This is a uniform and reproducible material that does not crack like the polystyrene. It does, however, have some rather serious mechanical instabilities, especially when temperature cycled.

PTFE/glass microfiber laminates were also introduced in the 1960s. The original material duplicated the dielectric constant of polyolefin (2.32) but had far superior dimensional stability and a much greater temperature range. A dielectric constant of 2.2 was achieved in the microfiber material by, once again, reducing the glass content. These two materials can be found today characterized as type GR material in MIL-P-13949.

A very popular dielectric material in the late 1960s was PPO (polyphenylene oxide). This material was touted as the ultimate in microwave laminates when it was first introduced. It had the ideal dielectric constant of 2.55, had very uniform and reproducible electrical properties, and was very stable both dimensionally and over temperature. Why isn't PPO the only material used microwaves today? PPO had a very large problem chemically: Not only would etching solutions make the edges of boards very sticky (something like a piece of pizza coming out of a pan if touched to some other surface), but it had a habit of developing fine cracks at unpredictable times after the material was attached to a case. These cracks were attributed to improper processing and usually occurred in the drilling process. The manufacturers of PPO specified that the material should not be punched but drilled. Most people adhered to this recommendation but still encountered problems. The reason was that in order to drill PPO you needed either a new drill bit or one that had just been sharpened. Anything less than that would be considered dull, and the properties of the material would be such that the bit acted as a punch and the fine cracks would show up at a later date. Usually these cracks occurred after the circuit had been tested and was well on its way to being shipped. Needless to say, PPO finally disappeared from the microwave laminate scene.

During the middle and late 1960s the high-purity (99.5 percent) alumina substrate material arrived on the microwave scene, and microstrip technology began its rapid rise to prominence in the microwave filed. Its high-dielectric constant (9.9–10) and smooth surface finish resulting in low-loss circuits made it an ideal material for high-density microstrip circuitry. Attempts were made in the 1960s to fabricate high-dielectric laminate that would compete with alumina substrates, but most were inconsistent and unreliable or had fabrication difficulties. A glass cloth loaded PTFE material with a dielectric constant of 6 was the only success in a long line of attempts. This material is still available today.

During the 1970s microwave laminates and substrates really came of age. This was due partly to the new technologies that made more consistent and reliable materials available and partly to the new systems being designed in industry that put pressure on laminate manufacturers to produce these materials.

Alumina substrates increased purity (99.6 and 99.7) during the 1970s and have finer particle size. Techniques for sputtering and deposition are refined so that thinner more consistent metallization is possible. At the same time laminates were moving in both directions regarding dielectric constant. They were going lower (2.17) for applications in the millimeter range and higher (10.2) to give the alumina market competition. The road to these new laminates was not smooth, however. The first high-dielectric material that was released worked rather well on the bench for breadboard or unstressed applications. (By *unstressed* we mean no extreme temperatures or humidity.) When subjected to these atmospheric factors there were great inconsistencies in the material. Also, many times the copper lines would lift off the material if a soldering iron was touched to them. All of these conditions, of course, were unsatisfactory and made the microwave industry wary of anyone who came to them with a high-dielectric microwave laminate. This persisted for a number of years, but, as in all things, time (and extensive research) heals all wounds. By the latter portion of the 1970s the material was reintroduced, and three separate manufacturers were now supplying it. This time it proved to be an excellent material and found many areas where it replaced alumina as well as finding many new areas of application.

As a result of the history of laminates and substrates briefly outlined above, the microwave industry has a broad cross-section of materials from which to choose. A listing of some of them is shown below:

Material	Dielectric Constant
Woven PTFE/glass	2.55
Woven PTFE/glass	2.45
Woven PTFE/glass	2.33
Woven PTFE/glass	2.17
Nonwoven PTFE/glass	2.33
Nonwoven PTFE/glass	2.20
Ceramic-filled PTFE/glass	6.0
Ceramic-filled PTFE/glass	10.2
Alumina	9.0 to 10

2.2 TEFLON[R] FIBERGLASS LAMINATES

Teflon[R] fiberglass laminates are probably the most widely used material in microwave today. Some may argue with that broad statement, but think of all the microwave circuits you have seen recently. With the exception of some microstrip circuits on alumina substrates, the circuits were on woven PTFE, microfiber PTFE, or high-dielectric ceramic loaded PTFE (E-10, Di clad 810, 6010). The microwave magazines present this same array of materials in their advertising. The Teflon[R] fiberglass laminate is the workhorse of the micro-wave industry.

In order to have a strong, consistent, and reliable workhorse, the materials are governed by a military specification. The one for microwave laminates is MIL-P-13949. As of this writing the specification is revision F, effective 10 March 1981. The specification has the following title: PLASTIC SHEET, LAMINATED, METAL CLAD (FOR PRINTED WIRING BOARDS), GENERAL SPECIFICATION FOR.

The specification becomes rather lengthy, and in an attempt to save you some time and frustration, we will present an outline of MIL-P-13949F showing the major headings. This outline breaks down the spec into its major headings. From these sections you should be able to find the area you need and look to any needed subsections.

MIL-P-13949F

1. *Scope*
1.1 Scope
1.2 Classification

2. *Applicable Documents*
2.1 Issue of documents
2.2 Other publications

3. *Requirements*
3.1 Specification sheets
3.2 Qualifications
3.3 Terms and definitions
3.4 Materials
3.5 Dimensions and tolerances of material
3.6 Prepreg characteristics
3.7 Characteristics of laminates
3.8 Marking of printed-wiring board materials
3.9 Workmanship of printed-wiring board materials

4. *Quality Assurance Provisions*
4.1 Responsibility of inspection
4.2 Classification of inspection
4.3 Material inspection
4.4 Inspection conditions
4.5 Qualification inspection
4.6 Quality conformance inspection
4.7 Methods of inspection
4.8 Test methods

5. *Packaging*
5.1 Preservation
5.2 Packing
5.3 Marking
5.4 General

6. *Notes*
6.1 Intended use
6.2 Ordering data
6.3 Qualification
6.4 Terms and definition
6.5 Solder float (specimen preparation)
6.6 Bow or twist
6.7 Lot identification number
6.8 Relative temperature resistance
6.9 Punching
6.10 Storage

This outline should, as previously mentioned, gives you an idea where to look for the information you need for your particular application.

One area of the specification that should be emphasized here is that of material designation. This area defined the material as base material, thickness, copper type and weight, thickness tolerance, and material grade. A typical designation is shown below. Each area of this designation will be investigated and explained according to MIL-P-13949F:

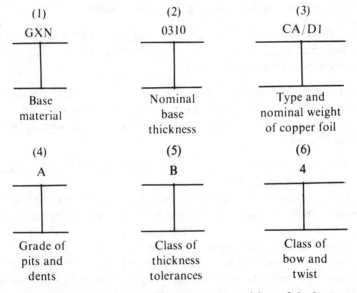

(1) *Base material:* This term defines the composition of the base material of the laminate. It contains three letters, but usually only the first two are used when speaking of laminate designation. Those letters will be taken from the list below:

- **PX:** Paper base, epoxy resin, flame resistant
- **GE:** Glass (woven-fabric) base, epoxy resin, general purpose
- **GF:** Glass (woven-fabric) base, epoxy resin, flame resistant
- **GB:** Glass (woven-fabric) base, epoxy resin, hot strength retention
- **GH:** Glass (woven-fabric) base, epoxy resin, hot strength retention, and flame resistant
- **GP:** Glass (nonwoven-fiber) base, polytetrafluorethylene resin, flame resistant

- GR: Glass (nonwoven-fiber) base, polytetrafluorethylene resin, flame resistant, for microwave application
- GT: Glass (woven-fabric) base, polytetrafluorethylene resin, flame resistant
- GX: Glass (woven-fabric) base, polytetrafluorethylene resin, flame resistant, for microwave application
- GI: Glass fabric base, polyimide resin, high temperature
- GY: Glass (woven-fabric) base, polytetrafluorethylene resin, flame resistant, for microwave application

The third letter in the designation can be:

- N: Base material without coloring agent or opacifier
- P: Base material with coloring agent or opacifier

(Unless otherwise specified N will be assumed if no third letter used.)

The base materials used in microwaves are GP, GR, GT, GX, and GY. They are characterized as follows:

- GT and GP: Suitable for loose tolerance designs or where circuits can be tuned, as well as for general-purpose printed circuit applications
- GX, GR, and GY: Designed for close tolerance applications of stripline and microstrip

Types GY, GT, and GX characterize the woven material and GP and GR characterize microfibers.

(2) *Nominal base thickness:* This a four-digit number used to identify the thickness of the laminate only. The number given in this example (0310) is equivalent to a laminate that is 0.031 in thick.

(3) *Type and nominal weight of copper foil:* This term completely describes the copper foil that is attached to the microwave laminate. The first two designators are for the cladding and weight of one side, while the second two are for cladding and weight of the other side.

Designators for cladding are:

- A: Rolled
- B: Rolled (treated)
- C: Drum side out, electrodeposited (ED)
- D: Drum side out (double treated), electrodeposited (ED)
- E: Matte side out, electrodeposited (ED)
- F: Matte side out (double treated), electrodeposited (ED)

Designators for weight are :

- R: 1/8 oz
- S: 1/4 oz
- T: 3/8 oz
- A: 1/2 oz
- M: 3/4 oz
- O: Unclad

(Any weight 1 oz or greater is designated as 1 oz = 1, 2 oz = 2, etc.)
Using the designators listed with the example shown we have:

- Side 1: Drum side out, electrodeposited, 1/2 oz copper
- Side 2: Drum side out, (double treated), electrodeposited, 1/2 oz copper

(Base material that is unclad on both sides would be designated 00/00.)

(4) *Grade of pits and dents:* This is a grading system used to determine the quality of the copper foil used on the finished copper-clad microwave laminate. It is graded A, B, or C depending on a point system that follows:

Longest Dimension (inch)	Point Value
0.005 to 0.010	1
0.011 to 0.020	2
0.021 to 0.030	4
0.031 to 0.040	7
Over 0.040	30

With this point system, the grades are determined as follows:

- *Grade A:* The total point count shall be less than 30 for any 12 in. x 12 in. area.
- *Grade B:* The total point count shall be less than 30 for any 12 in. x 12 in. area. There shall be no pits with the longest dimension greater than 0.015 in. Pits with the longest dimension greater than 0.005 in shall not exceed three in any square foot.
- *Grade C:* The total point count shall be less than 100 for any 12 in. x 12 in. area.

(5) *Class of thickness tolerance:* This term tells you how close the thickness is held to its specified value. Classes available are 1, 2, 3, or 4. The thicknesses with applicable tolerance given in the MIL-P-13949F specification are those that include the copper foil. When manufacturers specify laminates they do *not* include the copper foil. The thicknesses and tolerances shown for class 4 in the specification are:

Thickness	Tolerance
0.030 to 0.040	±0.002
0.041 to 0.065	±0.002
0.066 to 0.100	±0.003
0.101 to 0.140	±0.035
0.141 to 0.250	±0.004

(6) *Class of bow and twist:* The classes available for laminates are A, B, C, or X. Classes A and B are for standard manufactured-size sheets. Class B is a tighter specification than A. Class C is for laminated cut-to-size panels. For microwave applications (GP, GR, GT, GX, and GY) a maximum bow and twist of 3 percent is set for laminates of 0.030 in. or 0.031 in. The specification calls out 1–1.5 percent for material 0.060 in. and over. Class X is used to indicate no bow or twist requirements and may be used only for single-sided boards.

With the ground work now set for Teflon[R] fiberglass laminates by describing the specification that governs their fabrication, we are now ready to discuss specific laminates. Of the lower-dielectric laminate (2.45 and 2.55) there are two types in use today. They are the woven and the microfiber. The battle between these two has been raging since the microfiber material hit the market in 1960s. This text will not recommend one over the other because, quite honestly, there is no one best material when speaking of microwave laminates. The material used depends on the particular application. We will therefore present both types, indicate good and bad points of each, where applicable, and let you make a decision from the information presented. We will also present discussions on ceramic-filled Teflon$_R$ material that has a dielectric constant of 10 and competes with alumina substrates in many areas.

2.2.1 Woven Teflon[R] Fiberglass Material

This material will be covered first because it has seniority over the microfiber: It has been on the market since the mid-1950s. It was not the exact material then, however, that it is today. Its dielectric constant (ϵ) ranged from 2.65 to 2.75 as opposed to its present 2.45 to 2.55. You will recall from our previous history lesson how the original PTFE/glass (polytetrafluorethylene/glass) was developed for other uses but showed some promise for microwave circuitry, even though it proved to be a bit unruly at times. However, by the early 1960s the dielectric constant of the material had been stabilized and reduced to the now recognizable 2.45 and 2.55. The laminate was then ready to become a major part of the ever-growing microwave industry.

As the name implies, construction of this laminate is a woven pattern of fiberglass strands imbeded into the Teflon[R] material. If you were to look at your shirt under a microscope you would see a pattern very similar to that of the woven microwave laminate. A drawing of such a woven pattern is shown in Figure 2.10. The uniformity of this pattern is evident immediately. From this you can readily see how such parameters as dielectric constant and dissipation factor are also very uniform and consistent throughout the material. Figure 2.11(a) is an actual photograph of a piece of woven PTFE/glass laminate magnified 25 times. Once more, what stands out is the uniform construction as the fibers criss-cross through the Teflon[R] material. Figure 2.11(b) is the same material magnified many times.

TOP VIEW

SIDE VIEW

Figure 2.10 Woven glass laminate construction.

This uniform and consistent construction not only results in excellent consistency in parameters but also makes machining operations possible. This type of construction dictates that drilling operations should be accomplished with care. Carbide tip drills should be used with a slow feed to the material and a high speed on the drill press. The high speed is necessary, since the material is relatively soft and the hole is distorted if the drill speed is too slow. The same distortion results if the feed (both penetrating the material

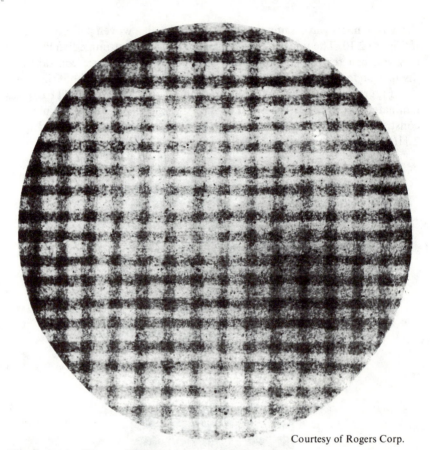

Courtesy of Rogers Corp.

Figure 2.11a Woven PTFE.

and backing out of it) is too fast. Of at least equal importance with the speed and feed of a drill bit used on PTFE/glass material is the bit itself. Solid carbide drill bits are the most useful and efficient, even though they need frequent sharpening or replacement. Qualities a good drill bit should have are a 60° included angle, 30° clearance, and a hi-helix tape. The first two requirements are shown in Figure 2.12. The third one is a definition of twist on the drill. There are three types available on drills: slow, standard, and fast. The slow is a very shallow twist on the bit; the standard is a general twist; and the fast is the hi-helix type, which is a fast or sharp-pitch twist in the drill construction. This hi-helix design is best for PTFE/glass drilling because it provides the maximum *shearing action* and rapid removal of drill shavings as it goes through the laminate. The woven material can also be sheared, cut, or milled to various shapes to conform to a particular shape needed for specific applications. One thing should be remembered after cutting the material, however. Since it is a woven-fabric type of construction there may be fabric edges left after machining. These edges should be removed with a knife or razor blade to ensure a smooth surface.

They can also be removed chemically with hydrofluoric acid. If an acid is used, however, the exposure time must be kept to a minimum, and the laminate should be rinsed quickly and thoroughly to prevent a wicking action. This wicking action occurs because of the construction of the material. The fabrics that are woven together act in the way a woven wick used in kerosene lamps does.

The tools to be used when cutting or shaping this type of laminate are very important, just as the drill bits are in drilling operations. Helical two-flute or four-flute end-mills are excellent for cutting PTFE/glass material. Also, the mills should be solid carbide, just as with the drills, because of their long cutting life. Speed and feed rate are also important when using the end-mill just as when drilling. Turning rates of 2,000 to 8,000 rpm and constant linear feed rates of 4 to 10 in per minute will result in excellent cutting and a smooth finished material. (Speed and feed rates depend on the material thickness and the diameter of the end-mills.)

Figure 2.11b Woven PTFE.

Courtesy of the MICA Corp.

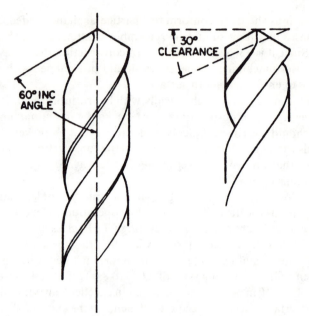

Figure 2.12 Drill bit for PTFE/glass.

As previously mentioned, woven PTFE/glass material started out with a dielectric constant of 2.65 to 2.75. In the early 1960s it was reduced to the popular 2.45 and 2.55. By varying the glass content of the material to an even lower percentage, dielectric constants of 2.33, 2.20, and 2.17 were put on the market in the latter part of the 1970s. This was a logical step to take when it is considered that the dielectric constant of pure Teflon[R] is in the area of 2.10. As you make the material closer to pure Teflon[R] the dielectric constant will be lower. Thus when the 2.55 PTFE/glass material had the glass content lowered to one point, a 2.33 or 2.20 dielectric constant resulted. Similarly, as the glass content was reduced even further the 2.17 dielectric constant resulted. This is about the lower limit with PTFE/glass laminates. Any further reduction in glass content will threaten the mechanical strength of the laminate. The 2.17 dielectric constant, however, is more than adequate for many applications in high microwave or the millimeter wave spectrum.

At the beginning of this chapter we covered four terms that characterized laminates: *dielectric constant, dissipation factor, dielectric thickness*, and *cladding*. These terms, and others of value, are presented in Table 2.2 for four different woven PTFE/glass laminates. These laminates have dielectric constants of 2.17, 2.20, 2.30, and 2.50 and are all commercially available. They can be ordered as standard products or as special if requirements differ from what is listed. These materials are available from three laminate manufacturers.

Table 2.2
Woven PTFE/Glass Laminates

Parameter	Laminate No. 1	Laminate No. 2
Dielectric constant	2.50 ± 0.04	2.33 ± 0.04
Dissipation factor	0.0018	0.0015
Thickness range	0.0035 in. to 0.250 in.	0.005 in. to 0.125 in.
Thickness tolerance	± 0.0005 in. to ± 0.006 in.	± 0.0005 in. to ± 0.005 in.
Cladding	ED/rolled	ED/rolled
Peel strength	12 lbs/in.	8 lbs/in.

Parameter	Laminate No. 3	Laminate No. 4
Dielectric constant	2.20 ± 0.02	2.17 ± 0.02
Dissipation factor	0.0012	0.0009
Thickness range	0.005 in. to 0.125 in.	0.010 in. to 0.120 in.
Thickness tolerance	± 0.005 in. to ± 0.005 in.	± 0.0006 in. to ± 0.0005 in.
Cladding	ED/rolled	ED/rolled
Peel strength	8 lbs/in.	8 lbs/in.

The cladding of each material is dependent on particular requirements and on individual manufacturers. The manufacturer usually will have both available but will classify one as standard and one as a special order. Check the data sheet of each manufacturer for its cladding specifications.

One term in Table 2.2 has not been explained: *peel strength*. This is the amount of force required, per inch, to separate the copper cladding from the laminate. This should be a reasonable number, since the copper must adhere to the laminate fairly well to have the circuit operate. Notice that as the glass content is lowered in the laminates the peel strength decreases. This is because more Teflon[R] is in the material and most materials do not adhere to predominately Teflon[R] surfaces very well. (Think of your fried eggs sliding around a Teflon[R] frying pan in the morning. Now picture these same eggs sticking in a variety of places where the pan has been scratched by metal utensils. This is analogous to the addition of glass to the laminate.)

Peel strength is also important when you are soldering to a copper-clad laminate. Material with a low peel strength will have a tendency to have the copper lift away from the laminate when heat is applied. This is a good reason to have a material that has a high peel strength.

Laminates are available that intentionally have a very low peel strength. These materials are used for breadboard applications that can be hand cut and peeled with a knife. These materials have a peel strength in the order of 4 lbs/in. and should be used only as a breadboard or for proving out a concept. DO NOT USE THESE MATERIALS FOR FINISHED CIRCUITS.

Another point to notice in Table 2.2. is the relationship between the dielectric constant and the dissipation factor. You will note that as the dielectric constant (ϵ) increases, the dissipation factor also increases. This is because the increased glass content in the material (to increase the dielectric constant) is considered to be an "impurity" added to the relatively low-loss pure Teflon[R] (PTFE) material. This increases the loss in the material, and, thus, a higher dissipation factor results. Increasing glass content no longer is feasible at the point where the losses become excessive at microwave frequencies. At this point the glass loading gives way to ceramic loading of the PTFE material. This combination gives results in a low-loss, high-dielectric (ϵ) microwave laminate. This material will be covered in detail in a later section of this chapter.

Since its release in the 1950s, the first "workhorse" laminate of microwave technology, the woven PTFE/glass laminate, has found many varied applications in microwaves. Today, as the microwave field expands to keep up with the fast-moving technologies, it still finds applications that make it an indispensable part of the huge and growing microwave industry.

2.2.2 Microfiber Teflon[R] Fiberglass Laminates

As previously mentioned, there are two types of Teflon[R] fiberglass laminates. The first, woven, was covered in the section above. The second, microfiber, first appeared on the scene in the 1960s. When this laminate is copper-clad, it conforms to MIL-P-13949 type GP and GR. You will remember from our discussions of MIL-P-13949 that GP was defined as glass-base (nonwoven) PTFE resin and GR as glass-base (nonwoven) PTFE resin for microwave applications. So the GP and GR designations readily define this nonwoven PTFE material.

We have seen what the woven material looks like and how it resembles the weave of a shirt. What, then, does microfiber PTFE/glass material look like? The official description of the material is a structure in which glass microfibers are individually encapsulated with PTFE (Teflon[R]) and evenly dispersed throughout the material with random orientation in the X,Y plane of the laminate sheet.

Actually, the material resembles dark mashed potatoes that have become hard and smooth. Figure 2.13 shows a picture of microfiber PTFE magnified 25 times. You will note its smooth consistent structure as compared to the woven material shown previously. This smoothness results in a laminate that has a dielectric constant that varies less than $\pm$ 1 percent and exhibits a low dissipation factor (0.0009 to 0.0012). The two microfiber materials available have dielectric constants of 2.20 and 2.33. Additional parameters are shown below:

Parameter	Laminate No. 1	Laminate No. 2
Dielectric constant	2.20 ±0.02	2.33 ±0.02
Dissipation factor	0.0009	0.0012
Thickness range	0.010 in. to 0.250 in.	0.010 in. to 0.250 in.
Thickness tolerance	±3 percent	±3 percent
	(±0.003 in. to ±0.0075 in.)	(±0.003 in. to ±0.0075 in.)
Cladding	ED/rolled	ED/rolled
Peel strength	15 lbs/in.	15 lbs/in.

Courtesy of Rogers Corp.

Figure 2.13 Microfiber PTFE.

This material is softer than the woven type and requires that certain precautions be taken. Excessive clamping pressure should be avoided; if clamping is necessary, use more clamps with less pressure. Also, be sure the work area is clean; material resting on a steel chip can, from its own weight, cause the the chip to become embedded in the material. Sawing should be done with skip tooth or coarse blade saws; a fine-tooth blade will tend to load up with material and will result in a rough edge. The material can be milled and drilled, just as its woven counterpart. Holes may become undersized, however, because of heat generated during drilling; these should be checked at room temperatures for their final size.

The material also has many advantages. Because of its structure, etching and plating solutions do not wick due to capillary action as they have a tendency to do with woven material. This same structure results in cleaner edges and holes when cut, milled, or drilled. It is also very stable dimensionally, being able to withstand temperatures up to 550°F (287.7°C) without warping.

To sum up microfiber PTFE/glass laminates we can say the following:

- Excellent dimensional stability;
- Uniform dielectric constant and dissipation factor;
- Material is soft and susceptible to chips becoming embedded in it;
- Clamping should be kept to a minimum;
- No wicking occurs; and
- Good processability.

With both woven and microfiber laminate covered in detail, this would be a good time to put to rest the argument about which material is best. We will present four materials, two woven and two microfiber. The 2.33 dielectric constants will be compared and the 2.20 dielectric constant will be compared. In this way we will not be comparing apples with oranges, which would result in useless conclusions:

	Microfiber **Laminate No. 1**	**Woven** **Laminate No. 1**
Dielectric constant	2.33 ± 0.04	2.33 ± 0.04
Dissipation factor	0.0012	0.0015
Thickness range	0.010 in. to 0.250 in.	0.005 in. to 0.125 in.
Peel strength	15 lbs/in.	8 lbs/in.
Water absorption	0.020 percent	0.024 percent
Surface resistance (megohms)	3×10^8	10^4
Cladding	ED/rolled	ED/rolled

	Microfiber Laminate No. 2	Woven Laminate No. 2
Dielectric constant	2.20 ± 0.02	2.20 ± 0.02
Dissipation factor	0.0009	0.0012
Thickness range	0.010 in. to 0.125 in.	0.010 in. to 0.120 in.
Peel strength	15 lbs/in.	8 lbs/in.
Water absorption	0.020 percent	0.020 percent
Surface resistance (megohms)	3×10^8	10^4
Cladding	ED/rolled	ED/rolled

You can see from this chart that there are very few differences in the materials. The argument that the microfiber material had less loss than the woven is invalidated by this chart. Actually, the dielectric constant of both materials is lowered by reducing the glass content. When this is done, the dissipation factor (loss) is reduced. This reduction will be essentially the same in both materials, as the chart shows.

Similarly, the argument that microfiber material is too soft can also be put to rest by discussing warpage of material. This is one test for hardness. Microfiber material can withstand up to 550°F (287.7°C) with no warpage. This should be rather conclusive evidence that the microfiber hardness should not be questioned.

In conclusion, we can say that all of the arguments presented are based on personal preference. The ultimate answer to the arguments about woven and microfiber PTFE/glass laminates is that the best material is the one that does the job for you. There is no one *best* laminate.

2.2.3 High-Dielectric PTFE

When we discussed PTFE/glass laminates in the previous section we talked about varying the glass content to adjust the dielectric constant. A decrease in glass meant a lower-dielectric constant and dissipation factor. Conversely, an increase in glass resulted in a higher-dielectric constant but also increased the dissipation factor. There was a limit where the dielectric constant was higher than standard material and dissipation factor was marginal (about $\epsilon = 2.7$). Unfortunately, this dielectric constant is not high enough for lower-frequency applications where the quarter-wavelengths become too long in standard PTFE/glass where $\epsilon = 2.55$. Alumina substrates, to be covered in the next section, with their dielectric constant in the area of 10, are excellent for the these applications. Unfortunately, they are difficult to machine, drill, and form to nonstandard sizes.

To fulfill the need discussed above, the PTFE laminates are *ceramic-filled* rather than glass-filled. The result is a Telflon^R board material with a high-dielectric constant, good loss characteristics, and flexible so as to be machined and drilled. This material is reference to in many ways: *high-dielectric Teflon^R, High-K material*, or *E-10* (for 3M's version of the ceramic-loaded material). Although it is one of the most revolutionary materials to come along in a long time, it was not always this accepted. When the material first came on the market in the 1970s, it presented many problems: Two prominent ones were water absorption and peel strength. Typical figures for the original ceramic-loaded material were 0.7 to 1 percent water absorption and a peel strength of less than 8 lbs/in.

Water absorption is a measure of the percentage of moisture absorbed by the material when it is soaked in water for 24 hours at a temperature of 23°C. The original numbers can be more fully appreciated when they are compared to woven and microfiber PTFE/glass:

Woven .... .01%
Microfiber .... .02%

You can see just how high a 0.7 to 1 percent absorption really is when compared to these materials. You do not want a material to absorb water, since it will disrupt such parameters as dielectric constant and dissipation factor. The latest data sheets for high-dielectric ceramic-loaded material show that materials dielectric constants of 10.2 and 6.0 have a water absorption of $<$ 0.1 percent (typically .05 percent). This material has come a long way since its introduction in the 1970s.

The second problem, peel strength, has also been improved. Typical peel strengths for the most widely used materials used today are shown below:

Laminate	Peel Strength (lbs/in)
Woven (ϵ = 2.50)	12
Woven (ϵ = 2.33)	8
Woven (ϵ = 2.20)	8
Woven (ϵ = 2.17)	8
Microfiber (ϵ = 2.33)	15
Microfiber (ϵ = 2.20)	15
Ceramic-loaded (ϵ = 6.0)	16
Ceramic-loaded (ϵ = 10.2)	16 (10 min)

The last item, ceramic-loaded (ϵ = 10.2), shows conclusively that the peel-strength problem has been resolved. Even the 10 lbs/in minimum figure is much better than the original value of less than 8 lbs/in. This improvement, and a general overall material improvement, has made the ceramic-loaded PTFE laminate a viable competitor for the well-established alumina substrate in the microwave area.

Typical specifications for the high-dielectric material available today are shown below. You will note that the specifications are given for both $\epsilon = 10.2$ and $\epsilon = 6.0$. This is because there are applications where a dielectric constant of 10.2 is too high and one of 2.50 is too low. This was why the 6.0 material was created. Thus we present both materials as high-dielectric laminates:

Parameter	Laminate No. 1	Laminate No. 2
Dielectric constant	10.2 ± 0.25	6.0 ± 0.2
Dissipation factor	0.002	0.002
Peel strength	10 lbs/in. min	16 lbs/in.
Water absorption	<0.1 percent	0.05 percent
Thickness	0.025 in. and 0.080 in.	0.020 in., 0.030 in., 0.050 in., and 0.062 in.

Both of the materials listed above are available with metal backing attached to them. Typical thicknesses of type 6061 roll mill grade aluminum are 0.0625 in, 0.125 in, and 0.1875 in. This is an excellent way of using this material because of the inherent softness of the material without a backing. This softness will cause a warping following a normal etching process. With an aluminum (or copper) backing this warpage does not exist. This backing of high-dielectric material is shown in Figure 2.14. This arrangement provides excellent dimensional and mechanical stability as well as guarantees a uniform electrical ground plane for your circuit.

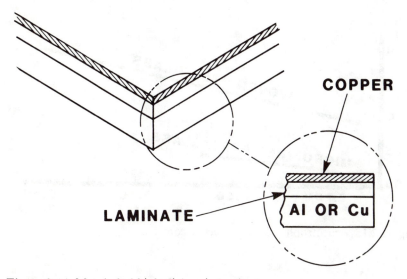

Figure 2.14 Metal-clad high-dielectric laminates.

The main point we mentioned earlier concerning the replacement of alumina substrates with a high-dielectric PTFE laminate was the machinability, or lack of it, of alumina. The laminates can be milled, drilled, or sheared similar to lower-dielectric PTFE materials. (Remember that this still is a Teflon[R]-base material, and proper ventilation is required.) It may also be sheared or cut with a razor blade or x-acto knife if only rough cuts are needed.

When connecting other circuits to those on high-dielectric material many conventional methods can be used. Everyday hand soldering methods are very acceptable and are frequently used. Thermocompression bonding has been done to some samples using a 300°C (572°F) hot stage. When solid copper cladding was used, the bonds were good. However when small pads or etched lines were used, they tended to lift from the substrate. This is because the thermocompression bonding hot stage softens the dielectric to a high degree. For this reason pulse or ultrasonic bonding should be used on these materials. (Bonding techniques will be covered in Chapter 6.)

High-dielectric PTFE laminates are finding more and more applications throughout the microwave industry. It should be remembered, however, that they are not the answer to every microwave application. Just as woven or microfiber laminates or alumina substrates have their specific places, so do high-dielectric laminates. It is up to you to find where they fit and use them appropriately.

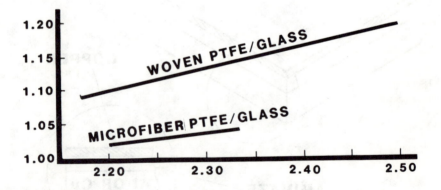

Figure 2.15 Anisotrophy versus dielectric constant.

Before moving on from soft substrates to the world of hard (or inorganic) materials, we should briefly look at the term dielectric *anisotropy*. This is a difference in dielectric constant in a microwave laminate in the X,Y plane of the material compared to that in the Z direction. This difference comes about because fill materials (ceramic, glass, etc.) are added to a pure material (PTFE) to obtain dimensional stability in the laminate. The laminate is no longer completely consistent in all directions when these fill materials are added. The difference in consistency is measured as anisotropy and is the anisotropy ratio, ϵ_{xy}/ϵ_z. Figure 2.15 shows this ratio as a function of dielectric constant for both woven and microfiber PTFE/glass material. You can see that the ratio ϵ_{xy} to ϵ_z is low (approaching 1.00) for low-dielectric constants and increases with dielectric constant increase. This is logical: Lower dielectrics have less fill material (glass) and are thus more consistent, since the laminate is almost a pure PTFE.

The figure appears to have a large distance between PTFE and microfiber but actually the materials are rather close. The ratios are as listed below:

Material	ϵ_{XY}/ϵ_Z
Microfiber $\epsilon = 2.20$	1.025
Microfiber $\epsilon = 2.33$	1.040
Woven $\epsilon = 2.17$	1.090
Woven $\epsilon = 2.45$	1.160

We should briefly explain why dielectric anisotropy should be a concern to the microwave designer. The major area of concern is in microstrip and stripline circuits, where it produces additional fringing capacitance. This additional component has the most prominent effect on resonator elements, in narrow lines, and in couplers using edge coupling techniques. In all of these cases the effects of fringing capacitance must be known and taken into account. If you have a material with a high anisotropic ratio and you do not know it, your circuits will not perform as expected. When choosing a material to be used in your microwave circuit, do not forget the term *anisotropic*.

2.3 ALUMINA SUBSTRATES

We have referred to alumina substrates on various occasions throughout this chapter. Now it is time to explain them and where they fit into the microwave industry.

Until the 1970s, when high-dielectric PTFE material was introduced, alumina substrates were the only reasonably priced materials on which microstrip circuits could be built. These substrates were referred to as the familiar alumina, aluminum oxide, or Al_2O_3. Their popularity began around the mid-1960s when their purity went from the previous low of 96 percent to a much more acceptable 99.5 percent. Since then it has gone to 99.7 percent and in some cases to 99.9 percent. The most widely used alumina substrates today use a purity of 99.6 or 99.7 percent. This increase in purity resulted in the "as-fired" surface finish's dropping from a previous 8–10 microinch thickness to approximately 4 microinches. The term *as-fired* means that this is the substrate after its final firing without any additional machining or polishing. You might call these "rough" substrates.

The two items that are of importance in an alumina substrate and make it useful for microwave application are dielectric constant and dissipation factor. The dielectric constant of the substrate is effected by two factors: the preferred orientation of the alumina crystallites and the density of the material. Fine-grained alumina ceramic has a more random crystalline orientation than one with coarse-grained properties. This random orientation is difficult to regulate and depends very highly on an accurate process control during manufacturing. This type structure, however, is still the preferred structure for the higher dielectric constant and lower dissipation factor. This is because the fine-grain structure results in a higher density in the material. The adhesion of circuit-forming films depends on the density of the grain boundary areas, and the fine-grain sizes produce a higher density that reduces *microwave scatter* caused by a coarse grain structure. This scatter is similar to a dissipation and increases the loss. It is preferred, therefore, to have alumina substrates with a fine-grain structure for the optimum dielectric constant and dissipation factor. This guarantees consistent substrate properties.

Other than the dielectric constant and dissipation factor, the flatness of a substrate is another important factor when choosing the right material for your application. Flatness is the term used to define the overall *camber* of a substrate. (Camber is likened to the warpage of a material; see Figure 2.16.) This is usually expressed in the allowable inch out of flat per inch of substrate (0.001 in/in). The term *camber* is important since flat substrates (or as near flat as possible) are necessary to ensure proper operation of vacuum fixture during exposure, a good fit in holders during processing, and a good mask contact during the printing on the metalization.

Besides *flatness*, other terms used to describe the final finish of a substrate are *as-fired* and *polished*. We have previously describes an as-fired substrate as the condition of a substrate after its final firing without any additional machining or polishing. You will recall that we called these "rough" substrates. These substrates, as might be expected, can have definite limits regarding flatness and dimension control. Most of the time these limits make the as-fired substrates unsuitable for microwave application. When this occurs you must specify *polished* substrates. These substrates offer excellent performance but usually are much higher in cost. These substrates are a fired material that is buffed smooth by mechanical grinding or lapping.

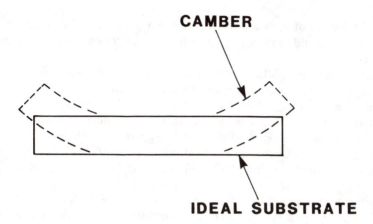

CAMBER

IDEAL SUBSTRATE

Figure 2.16 Substrate camber.

With the primary terms presented, we will close our discussions on alumina with a typical data sheet:

Material	Al_2O_3 (99.6 percent)
Dielectric contant	9.90 ± 2 percent
Dissipation factor	0.0001
Metallization	Chrome-gold
Surface finish	3 microinches (as-fired)
Camber	0.001 in. per in.
Thickness	0.025 in. $\pm$ 0.00005 in.
Outside dimensions	0.05 in. x 0.5 in. $\pm$ 0.002 in.

All of the terms in the data sheet above have been covered. This example should show you that alumina is an excellent substrate (electrically) for microwave applications. The only drawbacks in the material is its brittleness (it is rather like working with a dinner plate) and its difficulty to machine. Other than these factors, the substrate is an excellent choice for microwave usage.

2.4 SAPPHIRE, QUARTZ, AND BERYLLIA SUBSTRATES

These substrates probably should be put under a category of exotic. None of them could be called an everyday material that would be the first choice of a microwave designer.

Sapphire (also referred to as α alumina) is a material with a dielectric constant of approximately 11.0. This is about as close as we can pin this material down, since it is not isotropic and dielectric constants ranging from 9.3 to 11.7 have been measured on it. The difficulties in the fabrication of alumina (cutting, drilling, etc.) are also encountered with sapphire substrates.

The primary use of such a material is when a high thermal conductivity is desired. This would probably be the only real application of such a material for microwaves.

Quartz (SiO_2) is another brittle material that has the same fabrication difficulties as alumina and sapphire. Its dielectric constant is 3.78, which is somewhat higher than woven PTFE/glass material. This may be useful for higher-frequency applications where smaller wavelengths need a low-dielectric constant to keep the dimensions reasonable. The dissipation factor, unlike sapphire or alumina, is a relatively unimpressive 0.0015. This, you will recall, is equivalent to regular PTFE/glass laminates. Quartz would be used only in very special and rather loosely specified applications.

The last of the materials in this section is beryllia (BeO). It has a dielectric constant of 6.9 and a dissipation factor of 0.0002. This dielectric constant is put in the category of high-dielectric PTFE laminates ($\epsilon = 6.0$) and in the alumina category of dissipation factor. Beryllia substrates require polishing, since they have a coarse-grain structure resulting in an inherently rough surface and a poor resistance to chipping. These substrates are normally used only where an extremely high thermal conductivity is needed. This is because the material is extremely expensive and because of its high toxic hazard rating. This would normally be one of the last materials looked at for typical microwave applications.

2.5 CHAPTER SUMMARY

We have looked at a wide variety of microwave laminates and substrates in this chapter: the familiar PTFE/glass material, both woven and microfiber; the high-dielectric PTFE; alumina substrates; and sapphire, quartz, and beryllia materials. In addition to the materials, terms such as *dielectric constant, dissipation factor, cladding, metallization*, and *sputtering* have been covered to familiarize you with the terminology, or buzz words, of microwave laminates and substrates.

Microwave laminates and substrates are an integral part of the circuits they support. When you choose your laminate or substrate, take the time to investigate all of the possibilities and choose the right one wisely: Do not rely on what may be your "pet" laminate to do the job all of the time. Every microwave application is just enough different to justify a thorough investigation of laminates and substrates to come up with the one that will do the best job.

CHAPTER 3
METALS

3.1 INTRODUCTION

We have referred to metals in the first two chapters and will be referring to more later in the text. These metals are an integral part of the microwave circuits and systems in use today and will play an important role in those of the future.

Metals that find wide usage in microwaves are aluminum, copper, silver, gold, indium, tin, and lead. The last three are more prominent as combinations in solder than in a pure metallic form. These are not the only metals used, but they are the most prominent. You will recall from our previous discussions, for example, how chromium (chrome) and titanium tungsten were used on substrates as adhesive to keep gold attached to the ceramic substrate. These are not widely used metals but are important in specific applications. We will concentrate in this chapter on metals that have wide applications throughout the microwave industry.

The question that arises at this point is, What makes a metal suitable and acceptable for microwave usage? Two properties are of prime importance when considering a metal for microwave usage: (1) good electrical conductivity and (2) machinability.

Good electrical conductivity is important because the metal used in microwave circuits must carry high-frequency currents with as little loss as possible. This metal may be used as a conductor for a circuit (which is part of the ground plane). Regardless of the application, low loss (good conductivity) is necessary. Values of conductivity for the metals previously listed are given below:

Material	Conductivity (mho/meter)
Silver	6.30×10^7
Copper	5.85×10^7
Gold	4.25×10^7
Aluminum	3.50×10^7
Indium	1.11×10^7
Tin	0.877×10^7
Lead	0.456×10^7

The metals listed are in order of conduction: That is, silver is the best conductor, and lead is the worse of this group. This is not to say that lead is a poor conductor, only that it is poor compared to the rest of the metals listed. This, however, is not a problem, since it is not used for metallization on substrates or for cases that must be a part of a critical ground plane; it is used in solder with various other metals. Its conductivity by itself, therefore, is not a critical item.

Machinability of metals is important, since very few microwave systems will fit into standard square or rectangular spaces. Most have a cut out or an overhang or a curved portion needed to fit a rounded area. For this reason the metals must be sheared, drilled, milled, or generally shaped to conform to the necessary space. In order to be cost effective and efficient, the metals must be easily machinable.

We previously mentioned seven metals that find wide usage in microwaves. These will be covered in the following order:

- Aluminum
- Copper
- Silver
- Gold
- Miscellaneous metals
  - Indium
  - Tin
  - Lead

Before discussing each individual metal, it might be interesting to look at a periodic table of the elements and see where each of these metals falls. Figure 3.1 shows a table with the seven metals we will cover listed in their appropriate places. You can see that aluminum is by itself away from the other six. It is in the category classified as *very active metals*. The remaining six metals are in the category of *soft metals*. The metals silver (Ag) and gold (Au) are also listed in a separate group called the *noble metals*.

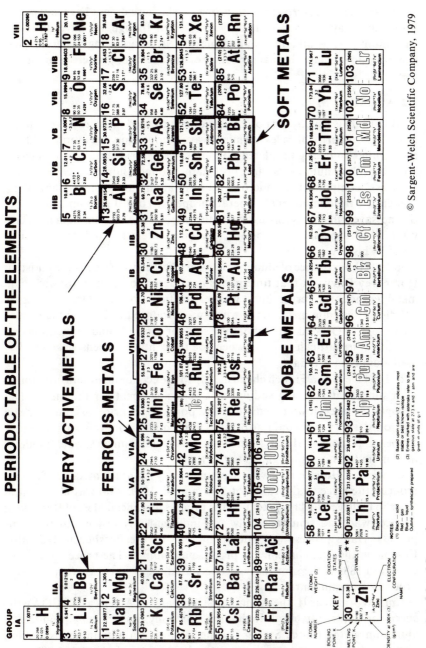

Figure 3.1 Metal Locations

The area termed *ferrous metals* is one in which we do not have any metals listed that may be used in microwaves. Even though we have discussed chromium (Cr) as an adhesive and nickel (Ni) as a plating material, these generally do not have the machinability or conductance properties of the soft metals.

3.2 ALUMINUM

Aluminum is usually first on the list of metals used in microwave applications: The majority of cases for microwave components are fabricated from aluminum. You see gold-plated aluminum cases, iridited aluminum cases, anodized aluminum cases, and just plain painted aluminum cases. Many "caseless" components in microwaves have aluminum ground-plane plates that enable the component to operate properly. The list goes on and on, but the message still comes through as to the importance of aluminum in microwaves. With a metal this important, it might be interesting to look into a little bit of its history.

In 1808 the existence of aluminum was predicted by the English scientist Sir Humphrey Davy. At that time it was considered to be the most abundant metal on the earth's crust, but it was very difficult to extract from the earth. The first aluminum metal in the world was produced in 1825 by the Danish scientist H.C. Oersted. At that point it sold for $160 a pound, something that may be rather difficult to grasp when you think how inexpensive and widely used aluminum is today.

In 1854 Deville, a French chemist, helped the economic situation of aluminum somewhat by using sodium as a reducing agent for the metal. This brought the price down to a mere $100 a pound. This price, of course, meant that only a very few could afford anything made from this special metal. Emperor Napoleon III of France had his finest dinner spoons made of aluminum to impress his guests in the 1860s.

Charles M. Hall found a way to inexpensively refine the metal in 1886 that reduced the price to the range of $8 a pound. Hall, who was a student at Oberlin College, filed his patent on the process just ahead of Paul Herould of France, who had simultaneously and independently discovered the same process. Hall's process was bought in 1888 by the Pittsburgh Reduction Company, the first company to produce aluminum. The process they purchased in 1888 is the same basic process used today in the current two-step process of aluminum refining.

Until World War II, the Aluminum Company of America (ALCOA), the successor to the Pittsburgh Reduction Company, was the only producer of aluminum in the United States. During the war, our aluminum production capacity was expanded greatly from the 1939 figure of about 164,000 tons. Although the majority of the plants built to handle this production were built and operated by ALCOA, their knowledge was made readily available to Reynolds Aluminum, the other prominent company.

The first cutback in aluminum production was made in December 1943, and by the end of 1944 the industry was operating at about one-half capacity. These cutbacks were needed simply because the supply of aluminum far exceeded the war needs. The huge industrial expansions that followed the war, however, more than made up for any cutbacks made during the war. The price of aluminum has decreased significantly even from the wartime prices. Following the war, in 1947, the price dipped to a record low of 15 cents per pound. The price, of course, did not hold at this level. Its present price of $2 to $3 a pound has been fairly constant for a number of years.

The only metal that has higher production quantities in the United States is steel. This is because of the wide application range and versatility of aluminum. Aluminum is a soft metal, although it is not listed as one in the periodic chart shown in Figure 3.1. You will recall that this group was called "very active metals." The metal is actually a soft metal that, almost contradicting logic, becomes so brittle at high temperatures that it can easily be powdered. The melting point of the metal is 660°C (1220°F), and it will boil at 2057°C (3734.6°F).

Aluminum is not the best conductor, but it has a very acceptable level of conductivity (3.50×10^7 mho/m as opposed to 6.3×10^7 mho/m for silver). Its softness enables it to be rolled into forms, cast, drawn, or stamped. It can, in its finished form, be milled, sheared, punched, or drilled. In short, it is a very machinable metal.

A general description of aluminum is given below:

Symbol	Al
Atomic weight (referenced to carbon 12)	26.98
Specific gravity	2.7 (a little more than 2-1/2 times as heavy as water)
Color	Silvery-white with a bluish tinge
Properties	Soft, easily shaped, resists corrosion, nonmagnetic, good conductor of heat and electricity
Chief ore	Bauxite (named after the town of LesBaux in Southern France where it was first mined)

This is aluminum: a soft and machinable metal with an average conductivity that finds applications in microwaves as cases and ground plates. The one property that accounts for its wide usage is its light weight.

3.3 COPPER

If the first metal you think of for use in microwaves is aluminum, the second would surely be copper. This is not a new metal by any stretch of the imagination: It was probably the first metal ever used by man. Objects of hammered copper 8,000 years old are known from Egypt, and specimens of cast copper from Egypt and Babylonia date from 4,000 B.C. The metal was given its name by the ancient Romans, who called it *aes cyprum* (metal of Cyprus), since the island of Cyprus was the chief source of copper for the Romans. Later they called it *cuprum*.

When the first Europeans came to the New World, they found the Indians using copper for jewelry and decorations. The copper these Indians used was from the Lake Superior area, deposited there by the action of the Ice Age. At that time, native copper was broken off from exposed lodes in the Great Lakes region and scattered southward over an area of over 70,000 square miles.

The mining of copper and the development of the industry began in the seventeenth century. In 1664 the first copper mine was established in the United States in Lynn, Massachusetts, and produced copper pins made from native copper in 1666. In 1780 Paul Revere established a foundry after discovering the secret process used by the English by which copper could be made malleable enough to be hammered when it was hot. The government loaned him $10,000 to buy a site for the foundry in Canton, Massachusetts.

In 1846 the first successful copper mine opened in the Lake Superior, Michigan, area. This is the time when the first gold mines were developed in Butte, Montana, which did not see copper mines until 1875. Mining was so extensive in this area that the entire city of Butte is undermined by the 700 miles of passages that run approximately 4,000 feet into the ground.

Copper is one of the metals in the category of coinage metals because it is used extensively in the manufacture of coins. There is a very high percentage of copper in coins today, much higher than in the past. Its price has risen sharply due to its scarcity, but this should not be a factor in influencing anyone to change his or her copper applications to another metal. The world is in no danger of running out of copper. Also, many applications do not require a pure 100 percent copper. Such alloys as brass or bronze do an acceptable job in many cases.

Copper fulfills our two initial requirements for a metal to be used in microwaves: (1) Its conduction is excellent (5.88×10^7 mho/m), second only to silver (6.3×10^7 mho/m); and (2) it can be drilled, tapped, sheared, sawed, and milled. One thing to remember when machining copper, however, is that special lubricants may be necessary, since copper is harder than metals such as aluminum. A quick check will save you from gathering a pile of broken taps and drills.

Copper is also one of the easiest metals to solder. It can also be heliarced, welded, or brazed. This ease of connection makes it ideal for low-loss microwave circuits.

Today, copper is not used as much as in the past because of its cost. If you looked in a waveguide catalog you would find extensive use being made of alloys of aluminum, magnesium, brass, and silver. These all make excellent conductors, are light-weight, and are lower-cost than copper. The laminate market, however, still uses 100 percent copper because it has found nothing to match copper's excellent conductive properties for the price.

A general description of copper is given below:

Symbol	Cu
Atomic weight (compared to carbon 12)	63.54
Specific gravity	8.9 (nearly nine times as heavy as water)
Color	Reddish brown (pink when freshly made)
Properties	Soft and easily shaped, becoming hard when cold-worked; good conductor of electricity and heat; resists corrosion by atmosphere and sea water
Chief ore	Chalcopyrite (a compound of copper, iron, and sulfur)

Copper, although not as widely used today as before, still has many applications in microwaves. Its contributions to the microwave laminate and substrate market are immeasurable, as you can see from our discussion in Chapter 2. Its versatility and properties should make it an important part of the microwave industry for many years to come.

3.4 SILVER

A second coinage metal is silver. This one is probably more recognizable than copper because most coins show the silver. Copper is usually visible only on the edges after extended usage wears the silver plating away.

Use of silver dates back to the earliest times. Ornaments of silver have been found in the Near East that date back to about 3,500 B.C. The book of Genesis mentions silver as part of Abraham's wealth. The first recorded mining of silver in Europe, however, was no earlier than 500 or 600 B.C.

Despite the demand for silver throughout the world, very little was produced during the Middle Ages because Europe's supply of silver was limited and its mining techniques were not highly developed. This shortage diminished when in the 1500s silver mines were discovered in Mexico, Peru, and Bolivia. Silver was not discovered in the United States until the 1700s.

During the Civil War, silver dollars and many other silver coins disappeared from circulation because the demands of industry for the metal caused it to be more valuable than the silver coins. After the war, in 1865, large silver mine discoveries in the Rocky Mountains decreased the price of silver.

Another silver shortage occurred in the early 1900s when large amounts were sent to India to help avoid a collapse of its currency system. A silver-purchase act passed in 1934 directed the Treasury to purchase silver, resulting in a large surplus. This surplus did not last, however, because ever-increasing uses were found for the metal. In July 1965 the U.S. government changed the metallic content of dimes, quarters, and half-dollars, and the result is the silver coin with the copper edges you see today.

The use of silver for microwave applications is almost entirely in plating. Waveguide and tuned cavities are made of a lightweight and inexpensive material (aluminum, etc.) and plated with a thin coating of silver to increase conduction of microwave energy. The coating is very thin, since the microwave energy travels only on the "skin" of the waveguide or cavity due to the phenomenon known as the "skin effect." It is much more economical and practical, therefore, to machine the cavity or fabricate the waveguide and plate it with the skin-depth thickness of silver. The price of silver is not quoted in this text because it changes from day to day.

A general description of silver is given below:

Symbol	Ag (from the Latin *argentium*)
Atomic weight (compared to carbon 12)	107.88
Specific gravity	10.49 (10-1/2 times as heavy as water)
Color	White
Properties	Soft and easily shaped; the best conductor of heat and electricity; resists corrosion by the atmosphere

3.5 GOLD

Most of the statements made in the previous section on silver could be repeated here for gold, starting by saying that its use dates back many thousands of years. The use of gold predates modern civilization, and many ornaments that have been found in Neolithic remains verify this statement.

A major use of gold in microwaves is for plating. Stripline and microstrip circuits are plated. Flanges and threads of microwave components are gold plated, as is the entire case or substrate carrier plate, to increase conduction and prevent corrosion of base metals used.

Gold is an excellent metal for use in microwaves but is being replaced — and in many areas has already been replaced — because of its price. For many years the price of gold was a nearly fixed figure, and any industry that had applications for it could plan its price and be assured of its accuracy. In recent years, however, the price of gold has varied from day to day (and sometimes hour to hour) and has jumped in price to a point where it is impractical in many cases to use it. Many circuits are now using tin/lead plating to save money, and connectors now use a stainless steel finish. These variations in plating have kept microwave components within a reasonable price range.

There are times when a gold plating is really the best metal to use for a particular application. When this is the case the designer usually makes the case in sections and gold-plates only the base plate, for example, which will act as the ground plane for the circuit. The remainder of the case is iridited, and the circuit usually operates very satisfactory.

A general description of gold is given below:

Symbol	Au (from the Latin *aurium*)
Atomic weight (compared to carbon 12)	196.96
Specific gravity	19.32 (nearly 20 times the weight of water)
Color	Golden yellow when pure; impurities cause various shades of yellow
Properties	Very soft and easily shaped; extremely resistant to corrosion; excellent conductor of heat and electricity; reacts to very few chemicals

3.6 MISCELLANEOUS METALS

Under this category of miscellaneous metals we will cover three metals that are in the "soft metal" group and whose primary use is in solders. The metals are indium, tin, and lead. Figure 3.2 shows the location of the three metals on the periodic table. You can see that they are all in a group and thus exhibit much the same properties. Each, however, has something that it does well.

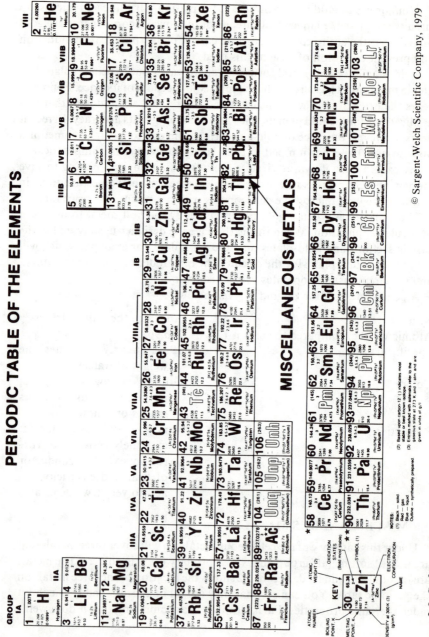

Figure 3.2 Soft metals

3.6.1 Indium

When you consider some of the metals we have discussed thus far, indium is a relative infant. It was discovered in 1863 by two German scientists, Reich and Richter. Indium is widely distributed in nature but occurs in commercially practical concentrations only in the ores of zinc, iron, lead, and copper. Because of this it did not have many applications but was a laboratory element. It was not until 1924 that Dr. William S. Murray decided to do something about the situation of indium to take it out of the laboratory and put it to work commercially. In 1926 Dr. Murray received a patent for his indium electroplating process. In 1934 he founded the Indium Corporation of America in Utica, New York, the world's first commercial producer of indium metal.

Indium has many properties that can be of great value in microwave application and in aiding the properties of other metals that could possibly be used in microwaves. Indium, for example, is so soft that you can scratch it with your fingernail. It will maintain this softness even at low temperatures. This is an excellent property if you are operating at low temperature and do not want your material to become brittle at these low temperatures.

Indium has no *memory*: That is, if you shape it one way, it will not return to the original shape but will stay the way it was shaped. One area where this property has found application is in using pill terminations in stripline circuits. The case in which the stripline is put will have tolerances on its milled portions. For this reason, some of the cases could put extra pressure on the termination, while others may just barely make ground contact. If you make the depression for the pill intentionally deep and put indium in it, the indium will fill the void, conform to the pill shape, make good electrical contact, and remain in that shape, even if the pill is removed from the case. There are other similar applications where indium could perform as well. Indium can be added to other metals to make them:

- Harder
- More fatigue resistant
- More ductile (hammered thin without breaking)
- Easier to melt
- Have higher heat conduction
- Have higher electrical conduction
- Easier to bond

These properties make indium a very useful metal. The improvements mentioned above when added to other metals make it an excellent choice for use as a solder for microwave circuits. A primary area where indium solders are used is when soldering to gold or silver. An indium/lead (In/Pb) solder will be an excellent choice over the conventional tin/lead (Sn/Pb) solder, since tin alloys soldered against either gold or silver tend to scavenge the gold or silver. (Scavenging is the attack of molten solder on a metal surface, with the solder actually dissolving the base.) The choice of using indium-based solder would be an excellent one if, for example, you were using alumina substrate with chrome/gold metallization. This would ensure that the solder joints would be solid, reliable, and exhibit high electrical conduction (something tin/lead solder probably would not do).

With all of the properties listed above, indium and indium solders should be a strong consideration for use in microwave systems. Be aware, however, that the metal also exhibits some properties that may be less than favorable. Reports have been filed that indium solder may develop long-term problems if all intermetallic interfaces are not considered. These problems seem more evident in space applications. With this in mind, be sure to investigate all of the conditions, both short-term and long-term, before choosing any solder or metal to be used for your application.

A general description of indium is given below:

Symbol	In
Atomic weight (compared to carbon 12)	114.82
Specific gravity	7.3 (about 7-1/2 times as heavy as water)
Color	Silvery white with brilliant metallic luster
Properties	Very soft material; stays soft at low temperatures; remains in shape when shaped; good conductor of heat and electricity

3.6.2 Tin

Most of the time tin is used either in a tin/lead solder or a tin/lead plating. About the only time you hear of tin used alone is in tin cans, and even these, curiously enough, are not pure tin but a thin tin plating on an inexpensive metal. Today many tin cans are not tin at all, and in the not-too-distant future any reference to a pure tin product may be a thing of the past.

Tin is another of the ancient metals that have been around for ages. Tin articles were known to exist at least as early as 1,400 B.C. in Egypt. Homer called tin *kassiteros* because he said it came from Cassiterides in the Atlantic. (The Cassiterides, or Tin Islands, were the British Isles.) The metal later got the name *stannum*, which resulted in its present symbol, Sn.

The only important ore of tin is cassiterite, or tinston, SnO_2, which is found in veins and streams, called alluvial deposits, mainly in Southeast Asia. (Bolivia and Africa also mine a certain amount of the metal.) Practically no tin is produced in the United States, which ranks first of users of this silvery-white metal.

As previously stated, the primary uses of tin in microwaves is in solder and for plating applications; the metal is excellent for both of these tasks. Remember, however, one precaution in the previous section about using tin/lead solder on gold.

A general description of tin is given below:

Symbol	Sn
Atomic weight (compared to carbon 12)	118.7
Specific gravity	7.29 (about 7-1/2 times as heavy as water)
Color	Silvery white with a bluish tinge
Properties	Soft and malleable; melts at a little over twice the boiling point of water; at very low temperature crumbles into a gray powder; conducts heat and electricity reasonable well

3.6.3 Lead

Lead, like tin, is one of the metals that do not appear alone in microwave applications. You will recall how we referred to tin/lead solder and tin/lead plating. Similarly, indium/lead (In/Pb) solder is an excellent substitute solder when gold is involved. Regardless of the combination, it seems that it always appears with another element.

Use of lead has its origins back many centuries, as do many metals. A small lead statue in the British Museum dates to the year 3,400 B.C., and lead is also mentioned in the Bible. In ancient days lead was sometimes confused with tin, and for this reason the Greeks used it very little. The Romans, however, used large quantities of what they called *plumbum*. Their main use was for water pipes. Actually it is the word *plumbum* that resulted in our word *plumbing* and in the symbol *Pb*.

The microwave applications, as previously mentioned, are in solders and in plating. These uses are somewhat limited but provide excellent materials for joining or protective plating microwave materials and components.

A general description of lead is given below:

Symbol	Pb (from the Latin *plumbum*)
Atomic weight (compared to carbon 12)	207.21
Specific gravity	11.3 (more than 11 times as heavy as water)
Color	Bluish-gray
Properties	Soft and easily shaped; resists corrosion by sea water, air, and many chemicals; fairly low melting point; the worst conductor of electricity compared to all metals covered (although not a bad conductor)

3.6.4 Kovar and Invar

Two other "metals" should be mentioned here, since they also find applications in microwaves where thermal considerations are important. These are kovar and invar. These alloys, which they are as opposed to pure metals, will be listed to be used as references later.

- *Kovar*: An iron/nickel/cobalt (Fe/Ni/Co) alloy with a coefficient of expansion similar to that of glass and silicon. Its thermal characteristics are similar to those of alumina substrates. It is used as a material for the mounting of alumina substrate to aluminum cases to compensate for differences in expansion, for headers, and in any glass-to-metal seals. (Notice that the three metals contained in this alloy are all in the ferrous metal group of the chart shown in Figure 3.1.) This metal can be difficult to machine.

- *Invar*: An alloy containing 63.8 percent iron, 36 percent nickel, and 0.2 percent carbon. It has a very low thermal coefficient of expansion and is used where thermal considerations are of prime importance and a minimum of machining is necessary. This material also can be difficult to machine.

3.7 CHAPTER SUMMARY

This chapter has covered seven metals and two alloys that play an important part in microwaves. You will recall that the two properties necessary for usage of a metal in microwaves were electrical conductivity and machinability. With each of the metals covered we stressed their conductivity, and each description listed as one of the properties that the metal was a soft metal. The only exceptions were the alloys, kovar and invar, which can be difficult to machine.

From the material covered in this chapter you should be able to understand why metals are an integral part of the microwave circuits and systems in use today and those of the future.

CHAPTER 4
MICROWAVE ARTWORK

4.1 INTRODUCTION

In order to make a shirt, you need a pattern. If you want to build a woodworking project, you have a set of plans to follow. If you were baking cookies, you would need cookie cutters to give you specific shapes. The list could go on and on, but the idea remains the same. If you are going to make an idea a reality, you need some sort of pattern with which to actually produce that idea.

The same holds true in microwaves. The circuits used for microwave applications are all etched from laminates or substrates and, as such, require a pattern: microwave artwork. The microwave artwork does not just happen. There is a definite progression that produces highly accurate and usable artwork. Basically, there are three steps to producing good microwave artwork: (1) the initial drawing; (2) the layout; and (3) the photo process. Each of these steps will be taken individually in this chapter and explained. Examples will be shown, and comments and suggestions will be inserted where appropriate to aid in producing the finest piece of microwave artwork possible.

Before we begin our discussion we should say that the artwork we are speaking of is the final film that is used to actually etch the microwave circuit. This may be a positive exposure, negative exposure, single-sided, two-sided with registration between the sides, or any other requirement your application may have. Regardless of the application, it is that thin piece of film that is our ultimate goal.

4.2 MICROWAVE DRAWINGS

The drawing for a microwave circuit is like the foundation of a building. If the foundation of that building is put together well, it will stand for years. Similarly, a good drawing will make the rest of the artwork process go much more smoothly and will result in a piece of artwork that can be used many times with a high degree of confidence.

When you sit down to begin a drawing for a microwave circuit, you should have a good idea of what its outside dimensions should be. These dimensions could be determined by a quarter-wavelength or the length of a particular line or any number of factors. They may even be dictated to you by a customer. However the dimensions are determined, this is the starting point.

Knowing what the outside dimensions are and the general layout of the circuit, the first item to consider is what scale you want to use: That is, How big do you want to make the drawing? If you have a circuit that is to be 0.5 in. by 0.5 in. you probably would want to make this drawing at least 10:1 so that the lines will be a reasonable size and you can dimension the circuit clearly. If, however, you have a circuit that is 3 in. by 3 in. you certainly would not want to draw that using a 10:1 scale. A 2:1 or 3:1 scale is appropriate for this application. How do you determine how large (what scale) you should make your drawings? As a good rule of thumb you should strive to have your finished drawing fill an 11 in. by 17 in. sheet. This rule will generally cover about every application you encounter in microwave circuitry. Any circuit that does not meet these requirements is a special case and should be handled accordingly.

One statement probably should be made at this point. The comments in this section are made with the assumption that you are not using any means of computer generation for your drawings but are making a circuit drawing by hand and then using some form of system to generate the artwork. We will cover a system in which the computer design you end up with in your CAD (computer-aided design) routines will automatically generate a drawing for you. For now, however, we will be dealing with hand-drawn drawings.

In order to demonstrate and illustrate our points on drawings, we will use a specific example: a two-way power divider that is shown in Figure 4.1. This is an ideal choice because it has both wide and narrow lines, has bends in the lines, and has components attached to the circuit (two chip resistors shown at A and B). The outside dimensions are set on this particular circuit and are 0.700 in. by 1.150 in. The scale that would be used on a 11 in. by 14 in. sheet of paper would be 10:1. This results in a circuit that is 7.0 in. by 11.5 in. This gives plenty of detail and allows you to dimension the drawing clearly. (Dimensions for this and all drawings throughout this text will be referenced to a 0,0 in the lower left corner.)

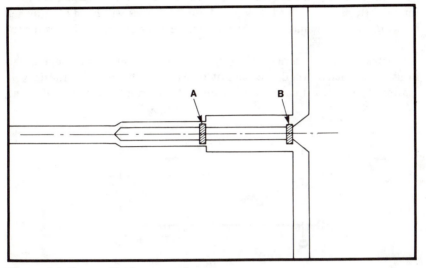

Figure 4.1 Power divider example.

The first step to take in making the circuit drawing is to find the center of the paper. This will ensure that the circuit is symmetrical and allow you to put connectors at the center of an edge rather than off to one side. The center, of course, is located 0.575 in. over and 0.350 in. up (Figure 4.2a). As we go through the drawing, the dimensions will be placed on the drawings as actual dimensions. That is, 0.046 in. for 50 Ω lines, etc. You must remember, however, that everything that you measure on the drawing must be 10 times these dimensions, since the scale is 10:1.

With the center of the paper established we can now put three 50 Ω lines on the drawing — the input at the left and two output lines on the right. The input line is exactly in the center of the finished circuit, so one-half of it will be on each side of the center line (Figure 4.2b). Since the 50 Ω lines are said to be 0.046 in. wide in Figure 4.1, we will need to add 0.023 in. to the 0.350 in. center dimension, draw a horizontal line 1, and then subtract 0.023 in. and draw a second horizontal line 2. These lines are shown in Figure 4.2b and should be drawn very lightly at this point, since we do not know exactly how long they will end up being.

The output 50 Ω lines have been moved 0.300 in. from the right side. This can be seen in Figure 4.2b. This allows connectors to be put on and still have some space to the end of the board. Usually SMA connectors are used on such a circuit, and their total flange dimension is around 0.560 in. Thus, half would be 0.280 in. We therefore allow 0.300 to account for tolerance on material and connectors. This 0.300 in. dimension is one side of the line. If we move 0.046 in. to the left, we will have the other side. Once again, be sure to draw these lines lightly as we do not know how long these lines will be in the finished drawing. Once they have been connected, we can darken in all lines for the circuit.

Our task now is to connect the three lines we have just drawn together to form the finished power divider circuit. One thing different about the lines we are about to draw is that not only is the width set but the length is also set.

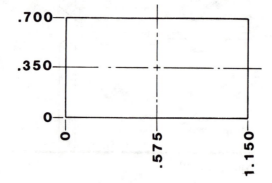

(a) LOCATION OF CENTER

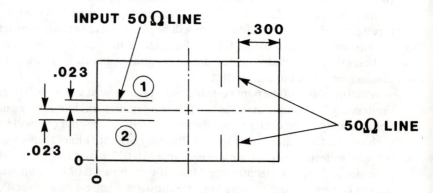

(b) 50 Ω LINE LOCATION

Figure 4.2 Power divider drawing.

To tie the divider together we will have to work from the right side.(output) lines, as we have said previously. The length and width of the lines connected to these lines are set and can be measured off rather easily. One nice thing about this circuit is that the top and bottom sections are mirror images. If you look back at Figure 4.1, you will note that everything above the center line ($\mathcal{C}$) is the mirror image of everything below. This can ease the drawing because you can figure out the top, for example, then fold the sheet and trace it for the bottom. Not all circuits, however, have this feature. Most of them require you to figure out and draw every line.

To continue with our drawing, refer to Figure 4.3a. This is what we have developed thus far. There is an input 50Ω line centered at the left and two output 50Ω lines 0.300 in. from the right-hand end of the circuit. We can proceed by noting in Figure 4.1 that a resistor is placed between these two output 50Ω lines. The dimensions of that resistor are 0.050 in. wide and 0.100 in. long. We therefore must have something less than 0.100 between the output lines so that we can solder the resistor to the circuit on both sides. In order to place it properly we will allow 0.084 in. for the resistor. This comes about from the fact that we would like to attach the resistor to the center of each line. Since the lines are 0.016 in. wide at the narrowest width, we will need 0.008 in. on each line or 0.016 in. total. If you subtract this from the 0.100 in. resistor measurement, you will have the required 0.084 in. dimension.

The next step is to move half down as shown in Figure 4.3b. This establishes one side of both high impedance lines and should be extended, very lightly, all the way to the input 50Ω line. We can now connect the output lines to the first high impedance line as shown in the expanded view. This is done by extending the 50Ω lines until they intersect the line just placed on the drawing (points A). Now measure 0.032 in. up from the line for the top portion and down for the lower portion. This establishes the first high impedance line width. Extend these lines through the 50Ω lines as you did before (points B). At this point you are ready to put the corners on the lines. This is done initially by connecting points A and B together as shown in Figure 4.3b. For optimum transmission of energy around the corner, some modification must be made. This modification is shown in Figure 4.4. The solid lines show a standard mitered corner where the 0.046 in. line and 0.032 in. line are extended and connected by line 1. The adjustment that is made is to move line 1 to line 2. This distance moved is called ΔW. As a rule of thumb we can consider this distance to be approximately 0.1 W (where W is the widest of the two lines being connected at a corner). The ΔW in our case, therefore, is 0.005 in., which is approximately 0.1 times the 0.046 in. dimension.

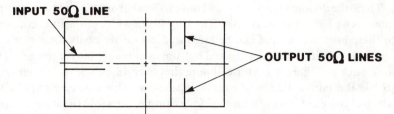

a) STARTING POINT FOR POWER DIVIDER COMPLETION

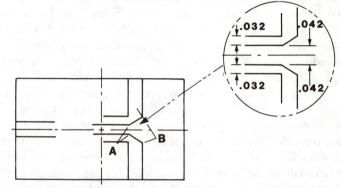

b) ATTACHING THE INPUT AND OUTPUT LINES

Figure 4.3 Power divider dimensions.

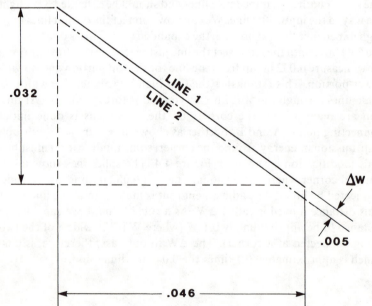

Figure 4.4 Mitered corner.

The final step in our construction process is to place the final high impedance line on the drawing and make the connection to the input 50 Ω line. We have already extended the lower line through the input line, so we need only measure up 0.016 in. for the top line and down 0.016 in. for the bottom line. The point as to where to place the line is determined by the length of the 0.032 in. line. Calculations show that this section of line must be 0.268 in. long. You therefore must measure from the center of the output lines to the left 0.268 in., and this will be where the 0.016 in. line will begin. This is shown in Figure 4.5.

Perhaps the most difficult part of this drawing is connecting the two 0.016 in. lines to the input 50 Ω line. This is because the lines have to be 0.268 in. long, just as the 0.032 in. lines were, but they also have to come together at the input line. If you are not careful, you can end up with a portion of the line that is wider than 0.016 in. and disrupt the impedance of this critical transition. Figure 4.6 illustrates this transition and how it is made. The key to drawing this section is 45° triangles and an ability to calculate the hypotenuse of these figures.

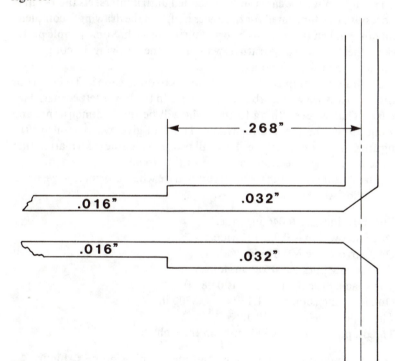

Figure 4.5 Line length determination.

In Figure 4.6a we see one-half of the transition to be made. If we start with the center line, we know that from the line to line A is a distance of 0.042 in. We also know that the line itself is 0.016 in. Since all dimensions for line lengths will be referenced to the center of the line, we will use one-half, or 0.008 in. of the line itself. The total vertical dimension is therefore 0.050 in. We can now measure 0.050 in. along the center line to the left. This is true because we are dealing with 45° triangles that have equal sides (isosceles triangle). Using the mathematical relationship that the hypotenuse of an isosceles triangle is equal to one side times $\sqrt{2}$, we have a length of 0.070 in. This means that 0.070 in. of the 0.268 in. length of the 0.016 in. line is in the transition. The remaining 0.198 in. is in the straight portion, as shown. This dimension is measured from the 0.032 in. line first, and then the angles are constructed as shown. To complete the transition measure 0.008 in. left and right of the 0.070 in. center line (points 1 and 2), and connect them to points 3 and 4, respectively. These should be drawn lightly until all lines are determined. The input 50Ω line can now be extended until it intersects the 1–3 line. This process is now duplicated for the lower half, and the drawing is complete. You can now darken the lines of the drawing if you wish. Some people prefer to darken them now, some rather wait until the drawing is completely dimensioned. The choice is yours.

With the drawing completed, our task now is to dimension it. To begin our dimension process we will establish a reference in the lower left corner. This will be 0,0. That is, everything to the right will be the X dimensions, and everything up will be the Y dimensions. This will give every point on the drawing an X and a Y coordinate. This will be very valuable information that will be used in the next section on layout of the circuit.

To begin our dimensioning we must summarize what information we have. This is as follows:

- The 50Ω lines are 0.046 in. wide.
- One narrow line is 0.032 in. wide.
- One narrow line is 0.016 in. wide.
- The narrow lines are 0.286 in. long.
- The spacing for the resistors is 0.084 in.
- Outside dimensions are 1.150 in. x 0.700 in.
- The scale to be used is 10:1.
- The output lines are 0.300 in. from the right side.

With this information we can begin the dimensioning process. Figure 4.7 shows the circuit prior to dimensioning. The only numbers on the drawing are the 0,0 datum point, the 1.150 in. length, and the 0.700 in. width. From this beginning and the information listed above the entire circuit can be dimensioned.

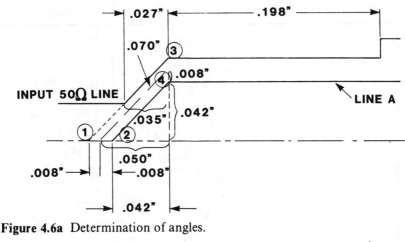

Figure 4.6a Determination of angles.

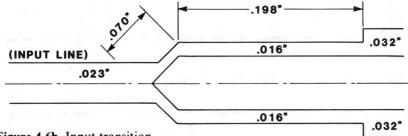

Figure 4.6b Input transition.

Figure 4.8 shows the circuit completely dimensioned. In order to understand where each of the numbers came from, we will take the dimensioning step by step to lead you to the final drawing of Figure 4.8.

- *Step 1:* The input line is centered. Therefore, each half must be dimensioned about the center line (0.350 in.). The dimensions are as shown (0.327 and 0.373), since the line is 0.046 in. wide and 0.023 in. is above the center line and 0.023 in. is below.

- *Step 2:* The output 50 Ω lines are 0.300 in. from the edge; therefore, one side of the line is 0.850 in. while the other is 0.804 in. There is no need to dimension both lines, since it should be obvious that they are on the same line. If, however, you think there might be some confusion, always dimension the lines. This will be the case even if you have to violate the rule set down by your high school or college mechanical drafting teacher who told you not to double-dimension a drawing. We are not after a passing grade here: We are after a drawing that will give a layout person all the information they need to make an accurate circuit.

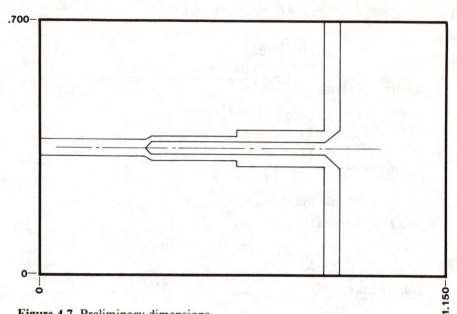

Figure 4.7 Preliminary dimensions.

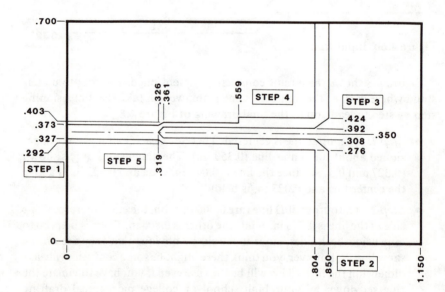

Figure 4.8 Power divider with dimensions.

- *Step 3:* The references for this step are the centerline and the requirement that there be 0.084 in. between the lines. If we start with the centerline (0.350 in.), we can go up 0.042 in. (0.392) and down 0.042 in. (0.308) and dimension these lines. The first line to the left of the output lines is 0.032 in. wide. The dimensions are therefore 0.424 in. for the top line and 0.276 in. for the bottom line.

- *Step 4:* To obtain this dimension we have to refer back to Figure 4.5. You will not that we measure the 0.268 dimension from the center of the output line. This starting dimension from this we will obtain the 0.559 in. dimension that is shown.

- *Step 5:* For this step we must refer to Figure 4.9. This is an expansion of the area referred to as Step 5. This is a good practice when you have an area on a drawing that could get cluttered or confusing when the dimensions are placed on the circuit. You usually can simply circle an area, put a capital A on it, and remark that the reader should refer to insert A.

  As a starting point to dimension this area we will take the 0.308 in and 0.392 in. dimension determined by step 3. The 0.016 in. lines are dimensioned as 0.308 in. and 0.408 in. from the step 3 references. The remainder of the dimensions come about from our discussions illustrated in Figure 4.6. You will recall that we set up 45° (isosceles) triangles to determine the position of the lines. These same triangles give us our dimensions. Since we know from Figure 4.6 that the lines are 0.198 in. long we subtract 0.198 from 0.559 and get the 0.361 in. dimension for the first line. We also know that the center of the intersection is 0.042 in. back and thus the 0.319 in. dimension. The final dimension, 0.326 in. comes about from the fact that this point is 0.035 in. back from the end of the line dimensioned 0.361 in. (This is shown in Figure 4.6.) By subtracting these numbers we obtain the 0.326 in. number.

With these steps followed we have now completed the drawing for our circuit. One further step will ensure that everything is complete. This step is to check that all of the lines are dimensioned. You can do this by taking a straightedge, starting at the bottom "0" line, and moving up the page. Every time you encounter a line you should have a dimension for it. If not, you need to put one on the drawing. Also, do the same by placing the straight-edge at the left "0" line and moving right. Once again, you should have a dimension for every line. Once you have checked this you are ready to send the circuit to layout.

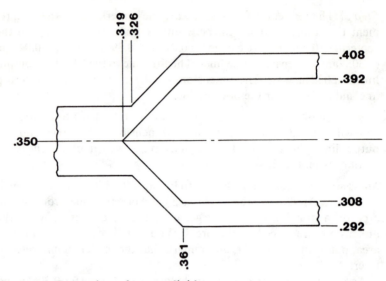

Figure 4.9 Expansion of power divider.

4.3 MICROWAVE LAYOUTS

In the previous section we stated that the drawing was like the foundation of a building. By now you should realize the truth in this statement. A similar importance can be placed on the layout of this drawing: If the drawing is the foundation, then the layout can be considered to be the walls. Just as a strong and reliable foundation is needed for a structure, similarly, it needs firm, straight walls that will add to that foundation. Both the drawing and layout are of great importance to the final microwave circuit.

There are many methods that have been used and are now being used to produce layouts of microwave circuit drawings. We will cover five methods that are in use today. They are:

- Printed circuit (PC) taping,
- Manual rubylith,
- Automated rubylith,
- CALMA,
- Autoart™.

4.3.1 PC Tape

The first stripline circuits were produced by taping the circuits on mylar using printed-circuit tapes. These circuits were usually laid out 10 or 20 times larger than required and then photographically reduced to obtain the needed accuracies. The tape used — and still used in some cases today — is called *pressure-sensitive graphic tape* or precision-slit PC network tape. It generally is a black tape but may also be a photographically opaque red transparent tape. The prime requirement is that it blocks the photographic light when the artwork is exposed.

This precision slit PC artwork tape is available in sizes ranging from 0.015 in. (0.381mm) to 2.0 in. (50.8mm), with some tapes available as wide as 6.0 in. (152.4mm). Some standard values available are:

Inches	mm
0.015	0.38
0.031	0.79
0.046	1.17
0.070	1.78
0.093	2.36
0.150	3.81
0.200	5.08
0.375	9.53
0.500	12.70
1.000	25.40
2.000	50.80

Tolerances on these tapes are generally good: For values from 0.015 in. to 0.312 in. the tolerances are ±0.002 in.; for values from 0.375 in. to 6.0 in. the tolerances are ±0.005 in. In millimeters this transforms as follows:

- 0.8mm to 8.0mm $\pm$ 0.05mm
- 8.5mm to 15cm $\pm$ 0.12mm

To layout artwork for a circuit, begin with the drawing made in the previous section. The first step is to set a border. In this case you do not want a full border but only corner markings that can be used in machining the outside dimensions of the circuit. These corner markings are shown in Figure 4.10. This is a taped layout of the power divider drawn in the previous section.

In order to make this layout, use the dimensions in the previous drawing. You are more concerned here with line widths and lengths rather than X,Y points as in the drawing: That is, we were concerned primarily with dimensioning all of the lines in the last drawing; now we are concerned with placing a piece of tape the actual width and/or length on the layout. As an example, the input line has dimensions of 0.327 in. and 0.373 in. at the left edge. We would thus measure up 0.327 in. from the lower edge and place 0.046 in. of tape at that point. The 0.373 in. dimension takes care of itself. This process is continued until the entire circuit is completed as shown in Figure 4.10.

As previously indicated the taping process was used for the first stripline, and some microstrip, circuits. In order to obtain the necessary accuracies the circuits had to be laid out on an expanded scale (X5, X10, etc.) and photographically reduced. This was an adequate method to produce artwork in the early stages of microwave circuit design. Most of the circuits in the 1950s did not require any amount of accuracy, mainly because designers knew that a high degree of accuracy was not possible and did not design for it. Some circuits are still laid out in this manner today. Usually they are done this way by smaller firms that do not have the capital to invest in sophisticated layout equipment and thus concentrate only on less critical circuits for their business. Also, some breadboard circuits use tape for the first layouts so that extra money will not be spent on an experimental design that may take many changes before the final circuit is achieved. Generally, however, the accuracy and smoothness of the line edges is not sufficient for most of the circuits required for today's microwave technology. For these reasons, other forms of microwave circuit layout were needed and found.

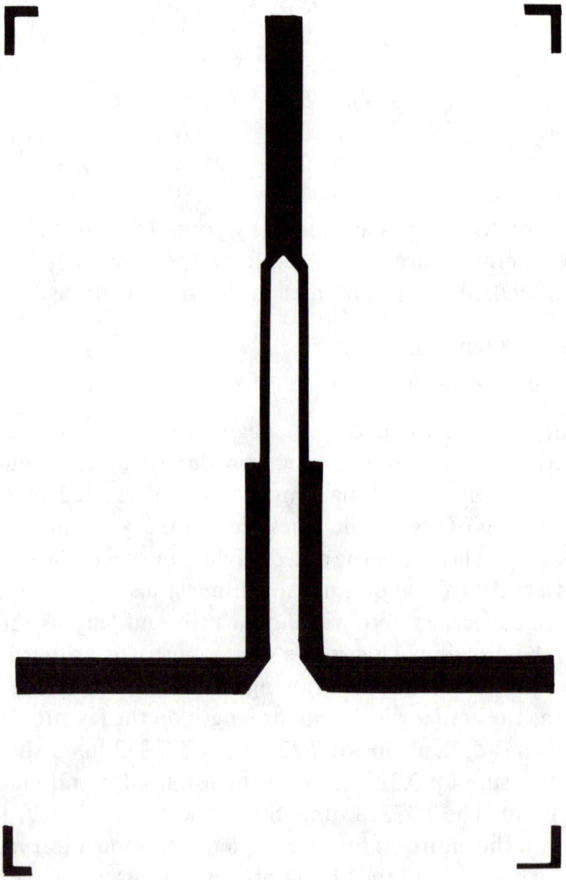

Figure 4.10 Taped layout.

4.3.2 Rubyliths

The term *rubylith* cannot be found in any dictionary, electronic or otherwise. It is a term used by the Ulano Company as a trade name for a photographic masking film. The name has been used so widely over the years that it has become *the* accepted term for this type of film throughout the industry.

The film consists of a mylar sheet covered with a peelable layer of soft transparent red material. Its construction is shown in Figure 4.11 with a picture of the material shown in Figure 4.12. The dimensions in parantheses are for another popular thickness of material. You can see that with these thicknesses the red film and mylar are of equal dimensions. This gives the film a good mylar base while producing a well-defined edge on the red film when cut. In Figure 4.12 you can see this edge on the lines and mitered corner that have been cut on this rubylith.

Figure 4.11 Rubylith construction.

To use the rubylith for creation of artwork you must cut out a particular circuit with a knife. There are two methods of accomplishing this: by tracing over a scaled-up drawing or by using an instrument called a coordinatograph. In the first method you could use the drawing generated in the previous section. This drawing would be to a certain scale (X5, X10, etc.) and would be placed under the uncut rubylith. This could be done on a drafting board or on a light table to allow the operator to see where to cut more easily. The circuit outline is now traced with the knife, and the circuit area is peeled off.

The coordinatograph is a precision drafting instrument that allows a high degree of accuracy in cutting artwork rubyliths. It consists of a large light table with a pen mounted on a highly accurate carrier. Many of these tables have a round indexed surface to allow for cutting of angles. This carrier also holds a blade arrangement that cuts the red film on the rubylith. The carrier is a screw-driven mechanism that is calibrated in both the X and Y directions. Artwork is cut in the coordinatograph by cranking in dimensions rather than tracing them as described earlier.

The drawing generated in Section 4.2 is once again used. The difference this time is that the 0,0 reference is established and all of the dimensions are put in with vernier scales that are highly accurate. This results in a much more accurate piece of artwork than would be produced by drawing on a light table.

To understand this consider that if we have a 10:1 layout on our original drawing, it is possible to trace the drawing and cut a rubylith within about 10 to 15 mils (0.010 in to 0.015 in). When reduced to actual size we are within 1 to 1.5 mils. This is not too bad, but consider further that the coordinatograph can layout this same 10:1 drawing within 1 to 1.5 mils originally. You can now see how much better the accuracy becomes when using this instrument.

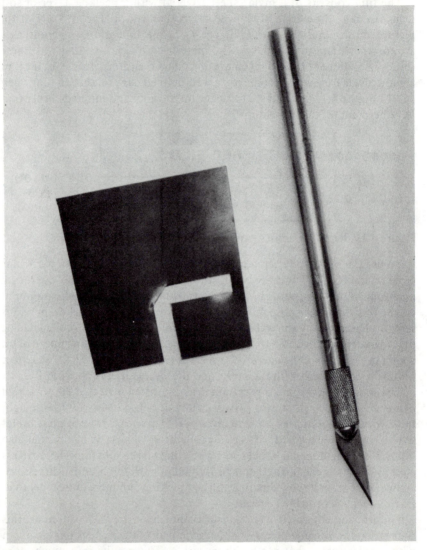

Figure 4.12 Rubylith material.

The rubylith is thus an excellent way to have a circuit laid out. This layout is now photographically reduced to produce the final sized artwork. It should be noted here that this photographic process must be very precise. That is, if a 10:1 reduction is called out, you should get exactly a 10:1 reduction; a 9.9 or 10.05 reduction will not give you the dimensions and thus the circuit performance you expect. The key word when cutting and using rubyliths is therefore *accuracy*.

4.3.3 Automated Rubylith

In the previous section we discussed the rubylith, its construction, and how it would be manually cut to produce a usable piece of artwork. This section discusses the same rubylith material but not manual cutting. Using a coordinatograph that is controlled by a computer results in an automatic rubylith cutting system.

The automatic coordinatograph is basically a motorized version of the manual device previously discussed. A controller is used to position the cutting unit and results in much greater accuracy, repeatability, and resolution than can be obtained manually. Additional advantages are:

1. It does not need the special environmental conditions many computer systems require and can thus be used in a standard office or drafting area.
2. It can be used as a stand-alone programmable system away from a master computer.
3. It can be operated by draftsmen or engineers easily because of its ease of programming and its ready availability in convenient working areas.

The artwork generated can be either a 1:1 reproduction or one that is scaled up. Figure 4.13 is a 1:1 reproduction of the power divider that we have been using as a model throughout the text. This layout was done on an Aristomat 100 system developed by AristoGraphic Corporation of West Germany. This system consits of:

- MAT 100 CNC controlled precision artwork cutting and measuring machine with plot-pen attachment;
- DKA-4 X,Y digital display;
- DSP-2 drafting machine controller with remote operator's console;
- GRE HA 3 tangentially controlled cutting device;
- DKA-4E expansion module incorporating a microprocessor;
- 100 PE dual floppy diskette module with alphanumeric CRT terminal; and
- Digitizer module.

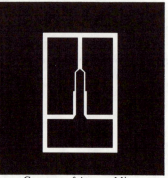

Courtesy of Anaren Microwave

Figure 4.13 Rubylith layout (1:1).

This 1:1 artwork could be used as the film to etch the power divider, or a scaled-up version may be cut and then photographically reduced. A 5:1 rubylith cut in the same system is shown in Figure 4.14. As previously stated, this rubylith can now be reduced photographically to a 1:1 piece of artwork and have improved accuracy of the dimensions, since the original accuracy at the time of the rubylith cutting is increased by a factor of 5.

As a means of checking the automatic rubylith system, a pen plot of what is to be cut can be printed. A plot for the power divider on a 5:1 scale is shown in Figure 4.15. This plot shows, first of all, whether anything is out or proportion on the circuit — for example, a line too wide or too narrow that would show up as you looked at the plot. Also, if two sides were supposed to be the same and they were not the same on the plot, there would be an error in the program that could be corrected before wasting a piece of rubylith by cutting the wrong circuit. The pen plot is a valuable piece of paper that should be checked carefully.

The automatic rubylith system is a large breakthrough for the microwave circuit designer. If offers many advantages over the manual method. Features of the automatic systems are:

- The systems (like the Aristomat 100 KE) are independent stand-alone precision-artwork cutting machines. They do not require a central computer for operation.
- Data can be created easily and typed in by an operator while the plotter is working. No computer programming skills are required.
- Programming is minimized by having repeat pattern capabilities within the system.
- Layout changes are performed with no loss of the original data.
- Artwork accuracy and machine repeatability eliminates rejections by quality control. This improves final parts accuracy, performance, and appearance.

- Use of smaller magnification scales for artwork compared to other methods results in less camera work, reductions, and step and repeat processes.

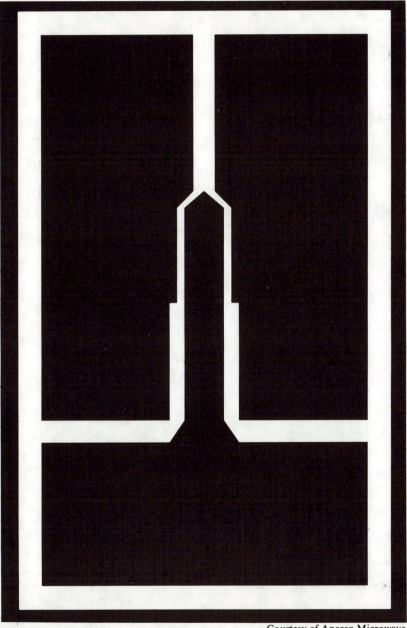

Figure 4.14 Rubylith layout (5:1).   Courtesy of Anaren Microwave

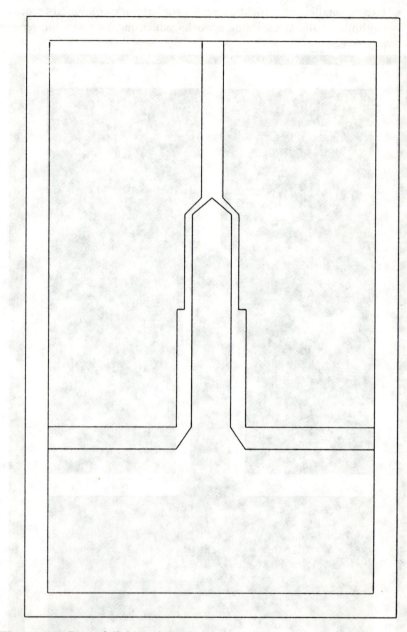

Figure 4.15 Plot of divider (5:1).

The automatic rubylith systems provide a highly accurate and repeatable piece of artwork of microwave circuits. This, of course, is an absolute necessity when dealing with stripline and microwave circuits with tolerances of ±.001 in.

4.3.4 CALMA Systems

CALMA is an interactive graphic system that can be used to lay out microwave circuits. By placing X,Y coordinates from our drawing into the system we can obtain a layout of the circuit in the form of a coordinate printout, pen plot of the circuit, and the complete circuit on tape that can be given to a plotter to make artwork.

The system consists of a display terminal that has a keyboard, two displays (one to show the layout and one to show the numerical information entered), and a tablet with an electronic tip for entering data points; the CPU (central processing unit), which is the main part of the system; a nine-track tape reader; and a printer. A basic block of a CALMA system is shown in Figure 4.16. A photograph of an actual setup is shown in Figure 4.17. The display terminal with two displays is shown on the far right; the CPU is on the far left; the nine-track tape unit is in the center right; and the printer is the center left between the CPU and the tape unit. Figure 4.18 is a close-up picture of the display terminal. You can see the keyboard and main display with the power divider on the left. The numerical inputs are displayed on the right screen with the tablet and pen just below the screen.

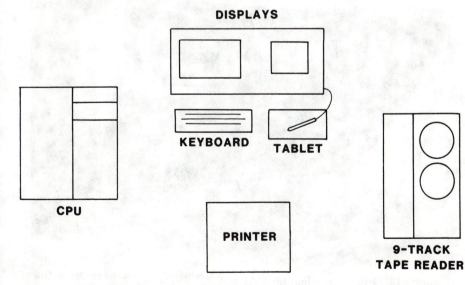

Figure 4.16 Block diagram of CALMA.

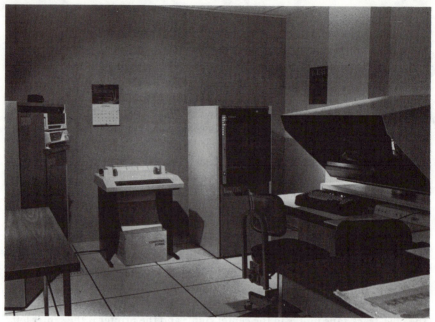

Courtesy of ITT Aerospace/Optical Div.

Figure 4.17 CALMA setup.

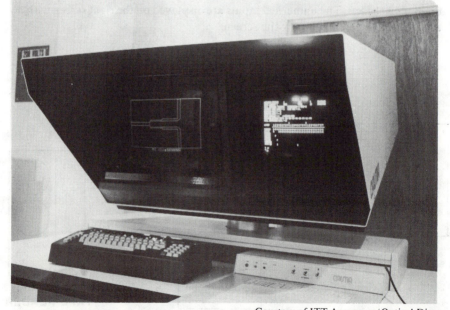

Courtesy of ITT Aerospace/Optical Div.

Figure 4.18 Display terminal for CALMA.

We noted above that our power divider was displayed on the terminal. This, of course, is an excellent picture to show if any obvious errors have been made. It does not however, give you anything concrete with which to check the actual circuit. What we can get from the CALMA system is a list of X,Y points and a pen plot of the circuit. Figure 4.19 is a list of X,Y points for our power divider. Record 1 shows the circuit outline, and Record 2 shows the divider. To show how this applies to our original drawing, refer to Figure 4.20. Point 1 shows up as X = 0.559 and Y = 0.424. This is shown back on Figure 4.19 as point 1. Similarly, point 2 shows X = 0.804 and Y = 0.276. Figure 4.19, point 2, shows this listing. You can start at the top of the chart and trace around the outline of the divider. This is a highly accurate check to find out whether what comes out of CALMA is what was put in. Figure 4.21 is the pen plot of the divider done on a Calcomp plotter. This serves the same function as the screen display: It shows up any obvious mistakes. It has the advantage, however, of being a permanent copy where as the screen is not.

Once the circuit has reached the point shown in Figures 4.19 and 4.21, it is ready for the final artwork. At this point it is on tape and is sent to a plotter to be outlined and transferred to photographic film. This can then be used to etch the required circuit. The process of etching will be covered in detail in Chapter 5.

4.3.5 Autoart™

Autoart™ is termed an interactive, two-dimensional drafting program for microwave circuits that translates the circuit descriptions into mask layouts. The program accepts a circuit description consisting of a list of interconnected elements similar to the X,Y printout we covered in the previous CALMA system. This nodal model is compatible with the microwave computer aided design program Super-Compact. Figure 4.22 shows a block diagram of Autoart and its interface with Super-Compact and other external accessories. You can see how the program can save much in the way of time and energy. When the circuit has been optimized, either by Super-Compact or a text editor, the program converts this circuit to a nodal model. Changes can be manually made to this model either by input commands at one stage (NODE mode) or by a graphics cursor at the second stage (GEOMETRIC mode). Once the design layout has been finalized, it can be passed directly to a graphics plotter or to a mask-making to produce final artwork. This artwork is produced on a 1.1 basis as opposed to some systems that cut the film 10 or 20 times and photographically reduce it.

```
CELL BEING ANALYZED IS 00000000


10/04/83      08:40:53




RECORD #      1
  X = 11500         Y = 7000
  X = 0             Y = 7000
  X = 0             Y = 0
  X = 11500         Y = 0
  X = 11500         Y = 7000

RECORD #      2
  X = 0             Y = 3270
  X = 3340          Y = 3270
  X = 3610          Y = 2920
  X = 5590          Y = 2920
  X = 5590          Y = 2760
  X = 8040          Y = 2760      POINT 2
  X = 8040          Y = 0
  X = 8500          Y = 0
  X = 8500          Y = 2760
  X = 8040          Y = 3080
  X = 3610          Y = 3080
  X = 3340          Y = 3500
  X = 3610          Y = 3920
  X = 8040          Y = 3920
  X = 8500          Y = 4240      POINT 1
  X = 8500          Y = 7000
  X = 8040          Y = 7000
  X = 8040          Y = 4240
  X = 5590          Y = 4240
  X = 5590          Y = 4080
  X = 3610          Y = 4080
  X = 3340          Y = 3730
  X = 0             Y = 3730
  X = 0             Y = 3270

RECORD #      3: CELL NAME = FLNOTE
                 :ORIENTATION IS, 0
                 :NODE, PR , JUST, R   LFT,  LOC. , 6000   0
                 :TEXT, MIC CELL:00000000


END OF FILE 00000000
```

Figure 4.19 X,Y points for power divider.

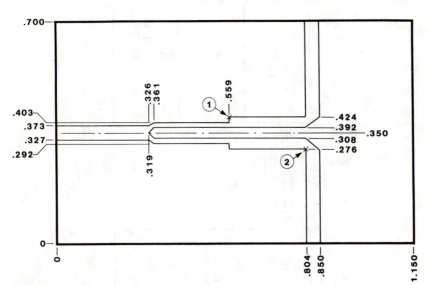

Figure 4.20 Power divider dimensions.

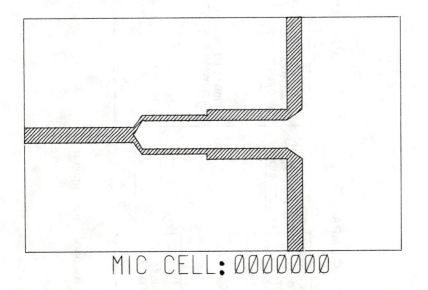

MIC CELL: 0000000

Figure 4.21 Pen plot of power divider.

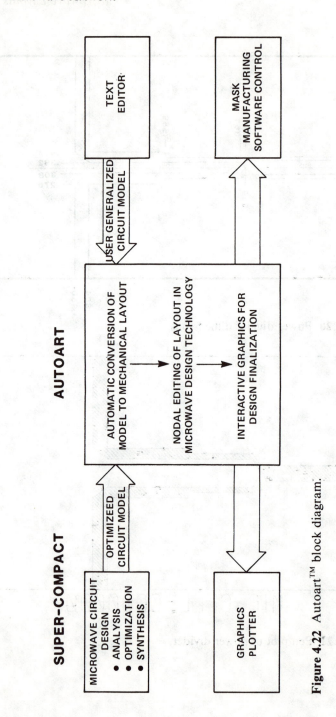

Figure 4.22 Autoart™ block diagram.

There are three *modes* of operation for Autoart. Some were covered briefly above when describing the operation of the program. The names and functions of these modes are:

- *AUTO mode*: Semiautomatic generation of physical MIC (microwave integrated circuit) layout from computer aided design file. (This is the final stage of the circuit layout.)
- *NODE mode*: Addition and manipulation of circuit entries. (This is where you can make any changes with input commands from a keyboard.)
- *GEOMETRIC mode*: Cursor-controlled movements of the geometric shapes on the screen. (This could be for more drastic changes made directly on the screen. These changes are then entered automatically by Autoart to update the data base of the layout).

Figure 4.23 shows a circuit alone on the Autoart NODE mode, and Figure 4.24 shows the final layout.

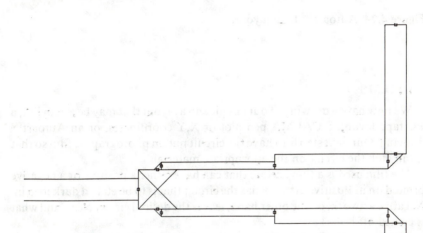

Figure 4.23 Autoart™ NODE mode.

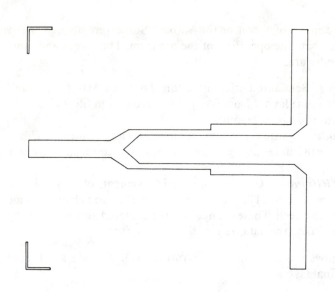

Figure 4.24 Autoart™ final layout.

4.4 FILMS

We now have a drawing of our circuit and a layout that may be a rubylith, a PC tape layout, a CALMA pen-plot or X,Y coordinates, or an Autoart™ drawing. Our next step is to have the circuit put on photographic film so that it can etch the circuit on the appropriate material.

The film used is a thin acetate that can have either a positive or a negative printed on it. Positive artwork has the circuit that is to be etched darkened in. Negative artwork, on the other hand, leaves the desired circuit clear, and what is etched off is dark.

When using the PC tape layout or the rubylith, the layout is photographed directly and a positive or negative is made from this. When the CALMA or Autoart™ is used, the coordinates are fed to a plotter system. One such system is the Gerber Plotting System manufactured by Gerber Scientific Instrument Company.

With most plotting systems you think of commands making a pen draw a certain shape. With the Gerber plotter a beam of light is shown onto the photographic film to trace out the outline of your circuit. The width of this beam, and thus the width of a line or gap between lines, is determined by the appropriately chosen aperture. The apertures are on a wheel that is inserted in the plotter. There are approximately 30 different apertures per wheel that can be chosen and a variety of wheels that can be purchased. As the tape control unit, using tape from CALMA or Autoart™, feeds coordinates to the plotter a *positive* piece of artwork is generated. The positive is generated directly, and a *negative* must go through the regular photographic process if it is desired. The artwork (positive) obtained from the plotter may be a 1:1 plot or may be an expanded layout that can be photographically reduced later.

When very narrow lines or very narrow gaps between lines are desired, or required, a glass mask is sometimes used instead of artwork made on acetate film. This type of mask is actually made out of glass with the required image on it and achieves a remarkable resolution.

It is one thing to produce a satisfactory piece of artwork for a circuit and another to keep it in good shape so that it can be used over and over again. Before leaving the topic of films we will present a series of do's and don'ts for handling and storing these films.

- *Do not* handle the films with your hands. Use gloves to keep from getting fingerprints on the film.
- *Do not* use disposable wipes to wrap your film. They contain glass fibers and can scratch the film.
- *Do* place your negatives or positives in photographic envelopes whenever possible.
- *Do not* expose the film to light — that is, *do not* leave them sitting on a desk or table for days.
- *Do* place nonabrasive material between films when they have to be stored in the same envelope.

4.5 CHAPTER SUMMARY

This chapter has presented the first step, beyond the design, toward the creation of microwave circuits. We have discussed the drawing of the circuit and how it is made; the layout, which includes PC tape, manual and automatic rubylith, CALMA, and Autoart™; and the films that record the final artwork. The work put into the artwork for a microwave circuit provides the pattern by which we create that circuit. Care should be taken to ensure that this pattern is made properly.

CHAPTER 5
ETCHING TECHNIQUES

5.1 INTRODUCTION

To this point we have discussed microwave materials (laminates and substrates); the metals involved with these materials and the cases that hold them; the proper way to make a microwave circuit drawing; five ways to layout this drawing; and the methods of obtaining final artwork (both positive and negative). We can now seriously consider etching our circuit.

The etching of a microwave circuit involves four basic steps:

- The cleaning of the material *thoroughly*;
- The application of the proper photo resist;
- The artwork placement and exposure of this artwork to identify the circuit area to be etched; and
- The actual chemical etching process.

In the basic steps shown above you will note that the word *thoroughly* is emphasized in the cleaning step. This is perhaps the most important step in the etching process. Many problems that arise in an etching process can be traced back to either an improper cleaning or the complete lack of it. To help avoid these problems we will present processes for cleaning substrates and laminates. There are times when these processes must be strictly adhered to, while other times only a basic cleaning is necessary.

97

All cleaning processes consist of three basic operations: (1) solvent bath, (2) rise, and (3) baking. The number of these steps may vary: More than one bath or one rinse may be involved, as is the case when the alumina substrates are cleaned to prepare them for metallization by the sputtering process. This process involves a solvent bath, an acid dip, a scrubbing, a detergent washing, three separate rinses, and a final baking at 1,100° C for eight hours. This is far from being a basic cleaning process.

A typical substrate or laminate pre-etch cleaning process is shown in block form in Figure 5.1 and illustrates the three basic operations previously mentioned in this block. The only deviation from this is a blow-dry process with nitrogen.

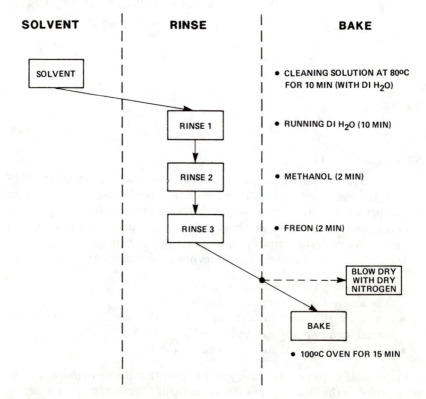

Figure 5.1 Substrate cleaning process.

Before individual steps in the cleaning process are explained, one point should be presented: Do *not* touch the substrate or laminate with your hands. Use nonmetallic tweezers or rubber finger cots. Fingers can leave a film of oil or dirt particles on the material that defeats everything you are trying to accomplish in a microwave cleaning process. When handling of laminates or substrates is required, use the proper precautions and save time and trouble.

If we now refer back to Figure 5.1, we can see each step and discuss why each would be used. To understand the steps of a cleaning process, it helps to know what is being cleaned off the laminate or substrate. Two basic types of contaminates must be removed in order to achieve a good etch: oil-based soils and water-soluble soils. Oil-based soils can be removed by vapor degreasing with trichlorethylene or an ultrasonic cleaning with trichlorethylene or freon. Ultrasonic cleaning is a process that vibrates a cleaning solution onto a material. In effect, the contamination is shaken off the material by the cleaning solution and the high-frequency vibration. Water-soluble soils are removed by using deionized water or high-purity alcohol.

Deionized water is defined as water that has been purified by removal of ionizable materials. Ionized materials are those that have electrons easily removed from electrons or molecules. This removal causes ions to be formed and an increase in free electrons that will reduce the electrical resistance of the material. For a thorough cleaning process, these materials should be removed from the cleaning solution.

One final comment before returning to Figure 5.1. The baking step is put into the cleaning process to remove any films that may be left from previous cleaning operations. This final step appears in all good cleaning processes; if it is not included, you should probably question the whole process.

The process shown in Figure 5.1 now should be much more understandable. The first step is a solvent bath with a cleaning solution (microcleaning solution, for example) that is a general cleaning process. You can then see that the three rinses are designed to remove both the oil-based soils (freon) and the water-soluble soils using DI (deionized) water and methanol. Blow drying with dry nitrogen avoids reintroducing the contaminates just removed, and the baking step removes the films left from previous processes.

In contrast to the process used to clean a substrate or laminate for etching, consider Figure 5.2, which is a cleaning process used to clean an alumina substrate in preparation for sputtering the metallization to it. It has many similarities to the process shown in Figure 5.1: cleaning with alcohol and deionized water, drying with nitrogen, and baking to remove any films resulting from the cleaning process. Additional steps taken in this process that were not used when preparing material for etching include a hydrochloric acid cleaning, an abrasive cleaning following a cold water rinse, a hot liquid detergent cleaning, and a total of eight hours of baking at $1,100°C$ as opposed to 15 minutes at $100°C$ for the pre-etch cleaning. Notice also that the rinses for the pre-etch cleaning are all to remove either an oil-based or water-soluble soil. In contrast, the premetallization rinses are primarily of hot or cold tap water to rinse a solvent from the substrate from a previous step. The only exception to this is the rinse with deionized water prior to the bake step. This rinse, as in other cleaning processes, is designed to remove any remaining water-soluble soils prior to this extensive baking cycle.

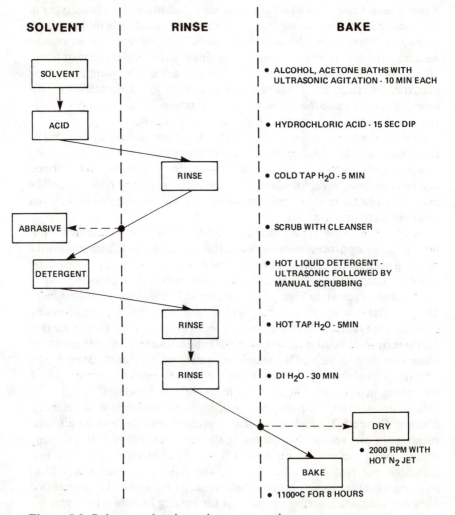

SOLVENT | **RINSE** | **BAKE**

SOLVENT — • ALCOHOL, ACETONE BATHS WITH ULTRASONIC AGITATION - 10 MIN EACH

ACID — • HYDROCHLORIC ACID - 15 SEC DIP

RINSE — • COLD TAP H_2O - 5 MIN

ABRASIVE — • SCRUB WITH CLEANSER

DETERGENT — • HOT LIQUID DETERGENT - ULTRASONIC FOLLOWED BY MANUAL SCRUBBING

RINSE — • HOT TAP H_2O - 5MIN

RINSE — • DI H_2O - 30 MIN

DRY — • 2000 RPM WITH HOT N_2 JET

BAKE — • 1100°C FOR 8 HOURS

Figure 5.2 Substrate cleaning prior to sputtering.

Following the cleaning processes the actual operations (metallization or etching) must take place. At this time the two processes listed vary completely. When metallizing an alumina substrate this substrate must be put into the deposition hot so as to prevent any recontamination of the material. This is vitally important since the adhesion of the metallization to the substrate depends very heavily on a clean uncontaminated surface.

Unlike the premetallization process, the pre-etch process requires that the substrate be cooled to room temperature before applying the photo resist. This ensures that the resist will adhere to the metal on the laminate or substrate so that a good image can be exposed.

We have emphasized the cleaning of substrates and laminates rather heavily in this section and for good reason. As indicated previously, many problems can be eliminated if the substrate is thoroughly cleaned before the photo resist is applied. The section on premetallization further emphasizes the importance of cleaning.

As a review of the cleaning process we will summarize do's and don't's as they pertain specifically to pre-etch cases: that is, how to clean a metallized substrate or copper-clad laminate prior to etching.

- *Do not* touch a metallized substrate or copper-clad laminate with your fingers. Use a set of nonmetallic tweezers or fingers cots. (If you would like to see what one finger can do to copper, touch a piece of copper-clad laminate with your thumb, and leave it on a table for a couple of days. You will see a dark outline of your thumb print on the copper that will be very difficult to remove.)
- *Do* use a liquid cleaning solvent. Avoid abrasive cleansers since they may scratch the metallic surface. These fine-grained scratches will cause problems with your microwave circuit (especially if 1/2-oz copper — or less — is used in your circuits).
- *Do* use a vapor degreasing with trichlorethylene or an ultrasonic cleaning with trichlor or freon for oil-based soils that may be present on the surface. Use deionized water or high-purity alcohol for water-soluble soils.
- *Always* end the cleaning process with a baking step that removes any films left from previous cleaning operations.

If these steps are followed, you should eliminate many of the problems that can arise during the etching process. Remember that the premetallized and pre-etch cleaning processes presented are only examples of what can be done to clean your substrate or laminate. Knowing what you are trying to clean and the degree of cleanliness required will dictate what your particular cleaning process will have for steps. Do not memorize, or copy word for word and step by step, what is presented here, but use the information presented and adapt it to your particular application.

5.2 PHOTO RESISTS

The process used for the etching of metallized substrates or copper-clad laminates is called *photolithography* (or photoetch). This process makes use of photographic techniques and materials called photosensitive polymers (or emulsions) called *photo resists*. The ISHM (International Society for Hybrid Microelectronics) Glossary of Terms defines a photo resist as "a photosensitive plastic coating material which, when exposed to ultraviolet light, becomes hardened and is resistant to etching solutions." This definition holds for both types of resists: positive and negative. This may be a bit difficult to grasp if you think of a resist put on a board, artwork placed on it, and the resist exposed to ultraviolet light. When a negative piece of artwork is exposed, the definition makes sense because the circuit you want is exposed and the resist hardens and resists etching. If, however, you use positive artwork, you would expose everything you do not want and the circuit would be etched away.

This would be the case if you took into account only that you apply resist, place the artwork on the material, expose it to ultraviolet light, and etch. There is one very important additional step: the step that develops the resist. This step distinguishes between processes. With this distinction in mind the following statements can be made concerning photo resists:

Negative: An emulsion that becomes insoluble when exposed to ultraviolet light. This makes the pattern you desire hard and resists the etching chemicals.

Figure 5.3 Negative artwork.

(Figure 5.3 shows negative artwork. You can see how the light will expose the pattern you desire and not the areas to be removed. Thus, the result on the material is exactly the opposite of the artwork: a *negative* image.)

Positive: An emulsion that is soluble in developing solvents. This resists duplicates the pattern you desire and thus the name *positive*.

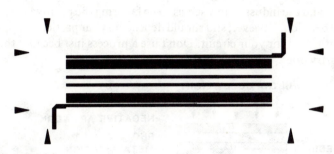

Figure 5.4 Positive artwork.

(Figure 5.4 shows positive artwork. All areas that will be etched away will be exposed to the ultraviolet light. The resist in these areas must be washed away by the developing process so that the only area with resist left on it is the area you want for your circuit.)

To understand further how the positive and negative processors work, refer to Figure 5.5. In both cases the substrate has a metallization on it. It may be 1/2-oz or 1-oz copper, a chrome/gold metallization, or a chrome/copper/gold combination. Regardless of its composition, there is a substrate with metal on it. The photo resist (positive or negative) is then applied to the metal (methods will be discussed later in this section). The artwork is then placed on top of the resist-treated substrate, and ultraviolet light projected down on the entire setup. In Figure 5.5a the light exposes only the desired trace. Following the developing and etching process the only thing left is that trace.

Conversely, Figure 5.5b shows that the ultraviolet light exposes everything *except* the desired trace. This allows all of this exposed area to be washed clear of photo resist in these areas and leaves it only over the desired trace. Once this process is completed and the etching is done, the result is shown as the same trace as in the negative process. We arrive at the same end using two different process: In one the resist stands up to developing and etching (negative), while in the other the resist is rinsed away where it is not needed (positive).

The decision as to which resist to use is comparable to the decision as to which material to use. There is no *best* resist, but there are applications where one may give better results than the other. Generally, if you have narrow lines to etch or have a narrow gap between lines, it is best to use a positive resist and, of course, positive artwork. This is because you can get a much sharper edge on a line or a gap that can be exposed to a higher degree.

Negative resist also has many applications. Up until the last few years, the largest majority of microwave circuits were etched using a negative process. Positive processes were used only in special cases. This has changed today, however, and the industry now seems to be favoring the positive scheme. The ultimate decision, however, still should depend on your particular application and what is best for your circuit. Don't use a process just because the rest of the industry does.

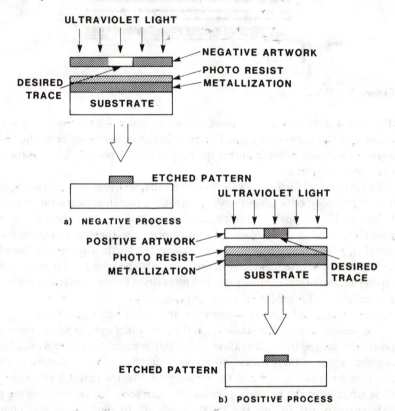

Figure 5.5 Positive and negative processes.

There are various ways to apply resist to the circuit board to be etched: by *immersion*, by *spraying*, or by *spin coating*.

The immersion technique consists of dropping the entire substrate into a resist bath in order to completely coat the unit. This can cause problems with woven PTFE/glass material if the resist gets on the laminate and "wicks" into the material a certain distance. This disturbs the dielectric properties of the material on the outside edges, which destroys the uniformity of the material. (Uniformity of material is necessary for proper microwave operations, as was stated in Chapter 2 on microwave substrate and laminates.) This process, therefore, is not a highly recommended one, although it is used in cases where it is the only method available.

In the spraying technique the substrate or laminate is sprayed with the resist, using either a spray can or other spray system. This process is much more acceptable, since it avoids the edge of the material and also controls the thickness of resist much better than immersion in a bath. Even though the control of coverage is much better than in the immersion method, there still is a danger of getting the resist too thick in one area or too thin in another area if a manual spray method is used. An automatic spray system provides a much better coverage over the entire substrate or laminate surface.

The process of resist application used most widely throughout the microwave industry is *spincoating*. In this process the substrate (or copper-clad laminate) is covered with the appropriate resist, and the excess is removed by spinning the substrate at a high RPM. The material is covered with a uniform coating of photo resist following this spinning, and the coating is then air dried and prebaked at a temperature that allows the solvent to evaporate without degrading the photosensitive properties of the resist.

To illustrate the photo-resist technique, we describe two processes: one for copper-clad laminates and the other for a metallized alumina substrate. Note how they follow the basic description presented above.

Process 1: Copper-Clad Laminates
- Following a thorough cleaning as described earlier (including a bake cycle), place the substrate on the proper-sized vacuum chuck using plastic tweezers.
- Apply enough resist to cover the substrate without having it run over the edges.
- Spin the substrate at 3,200 RPM for 20 seconds.
- After spinning allow the substrate to sit at room temperature for approximately five minutes.
- Place the substrate inside an oven at 90°C (194°F) for 20 minutes.
- Allow the substrate to cool to room temperature, then check the surface for foreign material or an uneven surface.
- If a ground plane is required for the back of the substrate, you can either put resist on the back side or cover the surface with mylar tape.

Process 2: Alumina Substrates
- Bake the metallized substrate at 125°C (257°F) for 30 minutes.
- Place the substrate in a vacuum chuck in the spinner, and cover the surface with photo resist.
- Spin the substrates for 15 to 20 seconds at approximately 4,000 RPM.
- Allow the substrate to air dry for 10 to 15 minutes.
- Bake in an oven at 90°C (194°F) for 25 minutes.
- If a ground plane is needed on the back, you can either spin a resist coating on the reverse side or put mylar tape over the metallization.

The processes shown above are typical processes for applying the photo resist. Although placing a coating on a metal-clad piece of material may seem elementary, it is one of the most important steps to be taken on the way to producing reliable microwave circuits.

5.3 ARTWORK PLACEMENT AND EXPOSURE

We now have cleaned our substrate and applied a thin uniform coating of photo resist. We now need to place the artwork on the substrate and expose it to ultraviolet light so that the required image is on the substrate.

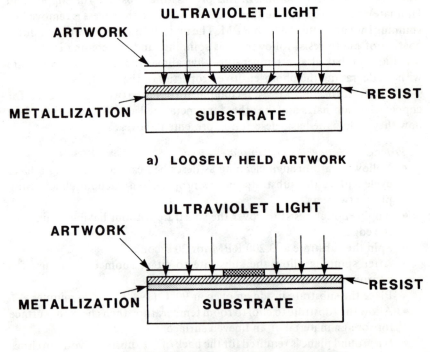

Figure 5.6 Artwork placement.

There is much more to artwork placement than simply laying a negative or positive on the material to be exposed. You will recall that all microwave circuits are based on specific length lines and, more important at this point, specific widths. Consider, for example, exposing a circuit board with a piece of artwork very loosely held to that board. Figure 5.6a shows this condition. You will not get good line definition because it is possible for light to get under the artwork, which will cause the width to be something other than what it was intended to be. This is similar to applying masking tape to an area that you are painting. You need a straight line for definition between two areas. The logical thing to do is put a piece of tape down to separate the areas, but if the tape is not put down firmly, some paint seeps under the tape to form a ragged edge.

When the artwork is pressed firmly to the material, as in Figure 5.6b, the light cannot get under the artwork. It exposes only what was intended to be exposed, producing good line definition that allows the circuit to perform as intended. Thus, a vital step in placement and exposure of artwork for microwave circuits is a vacuum that holds the artwork firmly in place.

The artwork (or mask) we have been referring to may be made of mylar or glass, as previously stated. Whichever type is used, the same importance is placed on having the artwork make firm tight contact with the material being exposed.

Generally, the procedure for aligning and exposing artwork (mask) is much the same from system to system. You should consult the manuals for your particular system for the proper settings to be used. Once you have these settings, the general procedure listed below can be used.

- Make initial settings on the alignment system as specified by the appropriate manuals (power, timer setting, etc.).
- Place artwork (mask) on the proper fixture with the emulsion side down. Do not touch the artwork with your fingers. This will leave fingerprints that may cause exposure problems. Use nonmetallic tweezers or finger cots. The emulsion side can be distinguished from the nonemulsion side because the emulsion side is dull and can be scratched. You will also notice that the circuit is raised somewhat on the emulsion when viewed through a microscope. (The fixture you use may be one supplied with the system or one fabricated specially for your application.) The emulsion is placed down so that you have the circuit impression in very close contact with the material you are attempting to expose. This ensures that if proper procedures are used, you will have the proper width lines exposed on the substrate. The film thickness with the emulsion side up could cause a variation in line width similar to that experienced by not having the artwork placed tightly against the substrate, as discussed previously.
- Place the fixture, with artwork, in the alignment system and apply the *minimum* vacuum. This is vitally important if you are using a glass mask, since too high a vacuum will shatter a $100 mask and upset a number of people. Begin with just enough vacuum to hold the fixture in place.

- Check for alignment of the mask on the substrate by viewing through a scope. Move to a proper alignment with either minimum or zero vacuum, if necessary. Alignment should be made with either corner or edge marks on the mask or artwork. (Figure 5.7 shows both edge marks and corner markings on a typical mask.)
- When alignment is completed, apply the appropriate vacuum. (Consult the system manual for the proper vacuum for each type of mask.) Push the table with the fixture, mask (artwork), and material into the machine. Previous settings will determine the exposure time. Typical times for frequently used resists are shown below:

Resist Type	Resist Name	Exposure Time
Negative	Kodak KTFR	15 sec
Negative	Kodak 752	15 sec
Positive	AZ 1350 J	10 sec
Positive	Hunts	8.5 sec
Positive	AZ 1400	10 sec

Table 5.1
Developer and Rinse Times

Photo-Resist Used	Developer		Rinse	
	Type	Time	Type	Time
Kodak KTFR (−)	Kodak Micro negative developer	≃ 2 min	Kodak Micro negative rinse	≃ 2 min
Kodak 752 (−)	Kodak Micro negative developer	≃ 2 min	Kodak Micro negative rinse	≃ 2 min
Shipley AZ 1350J (+)	Shipley AZD 135	15 to 20 sec	Di-water	≃ 1 min
Hunts HPR 204 (+)	Hunts Type-2 developer	15 to 20 sec	Di-water	≃ 1 min

- Following exposure, the table will either eject automatically or will need to be manually removed. When removed, turn off the vacuum and take the mask off the substrate. The material is now ready to be developed.

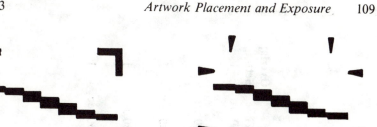

a) CORNER MARKINGS   b) EDGE MARKINGS

Figure 5.7 Alignment markings.

The developing process is designed to wash away the unexposed photo resist. There are two basic methods for developing microwave substrates: (1) immersion and (2) spin spray. The immersion is exactly what the name implies: developing and rinsing the substrate by immersing it in the proper solvents. A typical process is outlined below:

- Place the exposed substrate into a dipper, using nonmetallic tweezers, with the exposed side up.
- Immerse the substrate in the developer, and agitate for the time specified in Table 5.1.
- At the end of the required time, remove the dipper from the developer, and remove the excess developer by means of a blotting process.
- Immerse the substrate in the appropriate rinse, and agitate as shown in Table 5.1.
- Remove from the rinse, and blot any excess rinse from the substrate.
- Blow dry the substrate using dry nitrogen.
- Place the substrate in a 100°C oven for approximately 10 minutes.
- Following baking, the substrate should be checked for foreign materials, voids in the pattern, and the appropriate line width.

The spin-spray process is now outlined using the Solitex spin-spray machine.

- The initial procedure uses a test substrate with no solutions in the machine to set up RPM and time sequences. The DEV, RINSE, and SPIN buttons are depressed; the run switch is placed in the DRY RUN position; and the test substrate is placed in the vacuum chuck and the vacuum turned on.
- The START button is pushed and the DEV RPM is adjusted according to Table 5.2.

Table 5.2
Solitex Developer and Rinse Settings

Resist Type	Developer and Rinse Times		Solitex Settings						
	Developer	Rinse	Developer Speed	Time	Rinse Speed	Time	Dry Speed	Time	Overlap Setting
Shipley 1400	450 developer 2 parts to 1 part Di-water	Di-water	1500 RPM	30 sec	1500 RPM	30 sec	2000 RPM	30 sec	1
Shipley 1350J	Az developer no dillusion	Di-water	1500 RPM	30 sec	1500 RPM	30 sec	2000 RPM	30 sec	1
Hunt Positive	Type 2 developer no dillusion	Di-water	300 RPM	6 sec	300 RPM	30 sec	2000 RPM	30 sec	1
Kodak KTFR	Kodak (–) developer no dillusion	Kodak (–) rinse no dillusion	300 RPM	2 min	300 RPM	1 min	2000 RPM	30 sec	1
Kodak 752	Kodak (–) developer no dillusion	Kodak (–) rinse no dillusion	300 RPM	2 min	300 RPM	1 min	2000 RPM	30 sec	1

- When the machine cycles to the rinse mode, adjust the RINSE knob for the appropriate RPM shown in Table 5.2.
- When the machine cycles to the spin dry mode, adjust the SPIN DRY knob for the appropriate RPM shown in Table 5.2.
- The spin times are then set for DEV, RINSE, and DRY according to Table 5.2.
- The run switch is now put in the RUN position.
- The developer container is filled with the appropriate developer shown in Table 5.2.
- The rinse container is filled with the appropriate rinse as shown in Table 5.2.
- The system is now ready to develop the exposed substrate.
- Place the exposed substrate on the spinner chuck using nonmetallic tweezers or finger cots. Turn on the vacuum.
- The START switch is now pushed, and the system will automatically go through the sequence of DEV, RINSE, and DRY previously set with the test substrate.
- When the cycle is complete, release the vacuum and remove the substrate.
- Place the substrate in an oven at 120°C for PTFE/glass laminates and 150°C for metallized alumina substrates.

With either the immersion or spin-spray technique completed, the substrate or laminate is now ready for the step that will result in the printed microwave circuit: *etching*.

5.4 ETCHING

The etching process removes the unwanted metal from a substrate or laminate, resulting in the originally designed circuit. Metal can be removed in either of the two ways previously used for resist development: immersion or spray.

The most popular methods for etching stripline, and thus microwave, circuits for many years were the Kodak method and the Shipley method. These methods are also used to some extent today and are presented here in their entirety:

Kodak Method
1. Apply Kodak photo resist.
2. Expose copper-clad board through the negative artwork (mask).
3. Develop using Kodak developer.
4. Wash with MEK (methyl-ethyl-ketone) to remove the unchanged photo resist. (This step is not necessary for cladding on only one side.)

5. Apply Kodak dye, which will show up the exposed portions. (This step is optional.)
6. Etch with ferric chloride.
7. Wash with water to remove ferric chloride.
8. Wash with MEK to remove dye if used in step 4.
9. Pumice the circuit surface.

Note: If MEK is used it should not be allowed to remain on the material for any prolonged period of time because it will attack the laminate.

The Kodak method listed above is a *negative* process: That is, the artwork is negative; the resist is a negative resist; and the developer is a negative developer. When you see the artwork (or mask), your circuit will be the area that is light; any dark areas on the artwork will be etched away.

Shipley Method
1. Scrub the board with Shipley scrub cleaner 70.
2. Rinse with water.
3. Dry the substrate.
4. Coat with Shipley AZ-111 resist (AZ-1350 also can be used). Add a blue dye to the resist for better visual inspection.
5. Bake at least 20 minutes at 140°F (60°C).
6. Expose the board.
7. Develop using one part AZ-303 developer to four parts water. Developing time is one to three minutes. (If AZ-1350 resist is used, use AZD-135 developer.)
8. Rinse with water.
9. Air dry.
10. Etch with ferric chloride for approximately 5-1/2 minutes. This time could vary anywhere from one to five minutes depending on solution age, strength, and copper thickness.
11. Rinse with water.
12. Strip the photo resist.

This Shipley method is a *positive* process: That is, the artwork, resist, and developer are positive.

The methods shown above were designed to etch copper-clad laminates for stripline circuits. The etchant used — ferric chloride — resulted in excellent circuit boards; it has been used for many years to etch copper and is still widely used today. As previously mentioned, etching can be done by immersion or spray. Below is a list of some properties of a ferric chloride solution

that has special wetting, antifoam, and chelating (doping) additives to optimize etching performance:

Type	Immersion	Spray
Operating temperature	40° to 60°C	40° to 60°C
Etch rate @ 40°C	1 mil/min	0.5 mil/min
Time to etch		
1-oz copper	1.5 min	3 min

   You can see that the immersion method etches at a faster rate, 1 mil/minute
(1 mil = 0.001 in) and that it takes 1.5 minutes to etch 1-oz copper. This checks,
since 1-oz copper is 0.0014 in thick and therefore will take approximately 1.5
minutes to etch at a 1 mil/min rate. Similarly, the spray technique etching at
0.5 mil/min will take three minutes.

   Although ferric chloride has been the primary etchant for some time, other
types are used. These types are nonmetallic, ammoniacal, alkaline, copper
etching solutions. As soon as you begin to use these solutions there will be no
doubt in your mind that there is definitely ammonia in them. Their high
ammonia contact requires proper ventilation of the working area. (Vendor
data sheets should be consulted or the vendor contacted directly for proper
ventilation rates as well as other etchant precautions.) These etchants are
ideal for use when etching circuits on aluminum-backed high-dielectric laminates (E-10, for example) because ferric chloride causes a violent reaction
with aluminum and would completely destroy the aluminum plate if exposed
to the chemical. (If you have any doubts concerning this statement, get
yourself a bar of aluminum that is about 1/2 in square and 6 to 8 in long. Pour
a small amount of ferric chloride in a pan, and place the tip of the bar in the
solution. Be sure to have gloves on or to handle the bar with tweezers because
it will begin to bubble and disintegrate immediately upon touching the
solution and will become very warm over the entire length of the bar. If you
try this experiment, or have tried it, you will understand how valuable these
nonmetallic etchants are for this application.) (As an alternative to using a
different etchant, the aluminum plate can be coated with resist just as the
laminate is. This requires a more sophisticated resist application setup, but it
works very well.)

   A typical etchant that exhibits the nonmetallic ammoniacal properties is
produced by the MacDermil Corporation. It is a two-part etchant called
Metex that must be mixed at the time of etching. It is available in two forms:

- Metex MU-A, when used with MU-B, gives the most pH-stable etching
  solution.
- Metex MU-14, used also with MU-B, gives the most rapid etching
  solution.

This etchant must be mixed to form a final solution. The following procedure should be used to minimize the generation of ammonia vapors.

1. Add water to the tank (or container)
2. Add Metex etchant MU-B in the proper quantity.
3. Add the Metex etchant MU-A or MU-14 (whichever is chosen) below the solution surface.

(If addition of the MU-A or MU-14 is not possible below the solution surface, there will be a larger quantity of ammonia vapors that will have to be taken care of to protect the operator.)

A process using Metex MU-A and MU-B in an immersion method is shown below.

- Mix the etchant in an appropriate container as follows:
  - One part deionized water.
  - One part MU-B.
  - Two parts MU-A.
- Tape the reverse side of the substrate if the copper is to be left.
- Immerse the substrate in the etchant with the pattern side up. Agitate the solution across the substrate.
- As etching begins a blue film will flow from the substrate. When this film ceases, remove the substrate from the etchant and immerse in deionized water. (Time of etching should be checked carefully so that the circuit does not become overetched.)
- When the circuit is etched completely, rinse in deionized water for 10 minutes.
- Blow dry with nitrogen.

As discussed in Chapter 2, microwave substrates contain more than copper; they can contain chrome/gold and chrome/copper/gold. We have already covered the etching of copper and now turn to the etching of gold and chrome.

The first to be covered will be etchants for gold, since this is the first metal that must be removed from a multimetalized substrate. Etchants that can be used are one-part (transene type TFA, for example) or multipart (potassium iodine, iodine, and deionized water) solutions. Procedures for both are shown below.

One-Part (Transense-Type TFA)
- Under a ventilation hood, heat the gold etchant to 100°C. (Protect the back side of the substrate if a ground plane is required.)
- Place the substrate in a holder, and immerse in the etchant pattern side up. Agitate the substrate within the etchant.

- After a specified period of time (one minute, for example) check the pattern etching progress. If further etching is required, replace the substrate in the etchant for a few more seconds. Check until the required etching is complete.
- When the substrate is fully etched, place it in a rinse tank and rinse with running deionized water for approximately 10 minutes.
- Complete the process by blow drying with nitrogen.

(The process above was accomplished by immersion. The gold may also be etched using a spray etcher.)

Multipart Etchant
- Mix the etchant as follows:
  - 500 ml deionized water.
  - 100 grams potassium iodine.
  - 35 grams iodine.
  (Heat the mixture under a ventilation hood to 90°C.)
- Protect the back side of the substrate of a ground plane is required.
- Place the substrate in a holder, and immerse into the etchant pattern side up. Swiftly move the substrate through the etchant two or three times, and rinse with deionized water.
- If more etching is required, repeat the process. If not, remove the substrate from the etchant, and rinse it in running deionized water for 10 minutes.
- Complete the process by blow drying with nitrogen.

(Once again, this etchant may be used in a spray method. Etching time will be much faster than that of the one-part etchant.)

With the gold removed from the substrate you will have showing either a layer of copper (chrome/copper/gold substrate) or a chrome layer (chrome/gold substrate). If the copper is showing, use the process described previously for copper etching, which will reveal a chrome layer. If you are already at the chrome layer, proceed as follows.

Once again we will have two types of etchant: one-part and multipart.

One-Part (Transense-Type TFD)
- Under a proper ventilation hood heat the etchant (transene TFD) to 40°C.
- Protect the back side of the substrate if a ground plane is needed.
- Place the substrate to be etched in a holder, and place the unit in the etchant. Agitate the holder and substrate in the etchant.
- After 20 seconds remove the substrate, rinse with deionized water, and check for the degree of etching.

- If further etching is needed, place the unit back in the etchant for a few seconds, and check again.
- When the substrate is fully etched, place the substrate in a rinse tank with running deionized water for approximately 10 minutes.
- Blow dry the substrate with nitrogen.

Multipart Etchant
  - *Chromium Etchant*:
  - *Part A*: Potassium ferricyanide/100 grams to 3,000cc di-water.
  - *Part B*: Sodium hydroxide/500 grams to 1,000cc di-water.
- Mix the etchant three parts A to one part B. (Shelf life of this mixture is approximately two days.)
- Protect the back side of the substrate if a ground plane is required.
- Place the substrate in a holder, and immerse in the etchant and agitate. Check progress after 20 seconds by removing the substrate from the etchant and rinsing it with deionized water.
- If further etching is required, return the substrate to the etchant for a few seconds, and check again.
- When etching is complete (chrome is no longer visible around the gold), place the substrate in a rinse tank with running deionized water for approximately 10 minutes.
- Blow dry with nitrogen.

Once again, as in the case of the gold etchant, the solution can be applied with a spray etcher. Actually, the most convenient machine is one that can dispense all three etchants (gold, copper, and chrome) in the proper sequence, at the proper temperature, and for the proper time. Such machines are available and are tremendous cost and time savers.

One last step needs to be performed before checking all of the dimensions on an etched substrate and before checking for cracks in the lines or holes in the traces. That step is the removal (or stripping) of the resist that we took such great pains to put on the substrate a few pages ago. This is handled as follows:

- Under a hood for ventilation of the area, heat the appropriate resist stripper to 100°C.
- Immerse the substrate on a holder into the stripper with the pattern side up.
- After two minutes, check the pattern for resist by rinsing it with a spray of hot water to remove the suds.
- Any remaining resist may be removed by using a cotton swab dipped in stripper.

- Rinse the substrate in a hot water spray again, and place it in a rinse tank with running deionized water for approximately 10 minutes.
- Blow dry with nitrogen.

We now have our completed substrate, which we will now check for any defects. Problems areas that should be looked at are:

- Cracks in the lines;
- Pits in the metallization;
- Rounded corners on the lines instead of good mitered corners;
- Proper etching (good sharp edges on copper boards and only gold showing on multimetallized substrates);
- Proper width of lines (minimum amount of under cutting.)

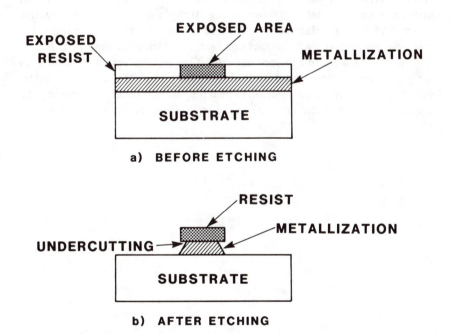

Figure 5.8 Undercutting.

You will note that the last statement says a *minimum* of undercutting. There is always a certain amount of undercutting of the metallization because it is not possible to have an etching process stop on demand. There is always a certain amount of creeping of fluid that will etch away some of the line beyond where you may want it to be. This undercutting is shown in Figure 5.8. You can see in Figure 5.8a how sharp the edges are on the exposed areas of the substrate. Figure 5.8b, however, reveals the real-world finished product, which is undercut. A good rule to go by when gauging etching procedures is that 1 mil undercut is acceptable. Anything larger than that would have to be considered regarding your particular application. If line widths are critical, anything over 1 mil (0.001 in) would have to be rejected. If your lines can tolerate 1.5 or 2 mils of undercut, then you can pass many more substrates through inspection.

5.5 CHAPTER SUMMARY

This chapter has taken us from the two-dimensional phase of microwave artwork to the three-dimensional real-world phase of a finished microwave circuit board. It emphasized the importance of cleaning material before an attempt is made to start an etching process. It explained positive and negative resists and how to place the artwork firmly on the material, expose the resist-coated material with ultraviolet light, and etch the circuit boards.

Producing a workable microwave circuit board involves much more than taping lines on a copper-clad laminate, placing it in a dish of ferric chloride, and shining a lamp on the solution to heat it up to etch faster. The etching of a highly sophisticated and accurate microwave circuit board is a science.

CHAPTER 6
BONDING TECHNIQUES

6.1 INTRODUCTION

Probably the most necessary, and least understood, process used in micro-wave circuit fabrication is that of bonding. This may consist of attaching components to a circuit, interconnecting substrates, or attaching substrates to a case. Regardless of the type of attachment, or bonding, referred to, its importance is often overlooked.

When we speak of bonding, just what are we referring to? Very basically, bonding is a method used to produce good electrical contacts between metal-lic parts (although some nonconductive bonding materials are used to attach microwave chips to a base substrate). It therefore requires methods that will join surfaces and produce good electrical contact. The methods investigated in this chapter include solder, epoxy, thermocompression bonding, thermo-sonic bonding, and ultrasonic bonding. Each of these methods has specific areas of application. Just as there is no one laminate appropriate for every application, there also is no one bonding method good for all applications. Each bonding requirement must be analyzed for such things as metallic content, environmental requirements, substrate content, and space require-ments, to name a few, to determine which is the best method of bonding surfaces. When the proper method is determined, the necessary operations must be performed to make this a proper bond both electrically and mechani-cally, something that is sometimes more difficult than choosing the proper type.

The following sections discuss the various properties of solder, epoxy, and bonding and then explain how these techniques are executed. These discussions will help you choose and then perform an attachment with the highest efficiency.

Before proceeding to the individual methods for microwave bonding we should bring up a very important subject: cleaning. It is imperative that all surfaces that are to be joined be as clean as possible to ensure a connection that not only holds initially but keeps on holding for the lifetime of the component. Consider what would happen if you dropped a ceramic statue on the sidewalk and broke it and one piece dropped into a pile of dirt and was covered with soil. If you put glue on the broken pieces without cleaning them, they would not hold together. Similarly, if there is dirt or contamination or oil on the surface you wish to solder to, or epoxy to, or bond to, you can be just as sure that you will not have a good connection. Be sure both surfaces are clean prior to attempting to bond them together.

6.2 SOLDER

The term *solder*, unfortunately, is one that practically everyone recognizes, and its familiarity allows for predetermined ideas about what solder is and how it is used. So many people think of the soldering process as one in which an iron or torch is used to heat the area to be joined and solder is fed to that area. The general consensus often is that the more solder put on, the better the connection: As long as the solder keeps melting, keep feeding it to the joint. This usually gives a connection, but its reliability can be debated. The difference between a soldering method and the *right* soldering method is one that should be clarified.

Figure 6.1 shows the requirements needed for the right solder connection. We will investigate each of these requirements, providing you with the information needed to make the decision about which method is best for your application

Our discussion begins with a definition of the term *solder*. (We will be discussing both tin/lead and indium-based solder.) It is generally defined as a meltable metal or alloy that joins two metal surfaces. The melting point of solder, of course, must be lower than the metals it is joining. The solders used for microwave application are termed *soft solders* because of their low melting points: below 450°C (842°F) and usually are well below 300°C (572°F). Common ordinary tin/lead (SnPb) solder melts at 190°C (375°F), and most indium solders, which are common in microwave circuits, generally melt at 150°C (300°F) or lower. There are also *hard solders* that have melting points above 450°C; these obviously are not used for microwave circuits, since they would cause destruction of the components being attached long before the solder even began to melt.

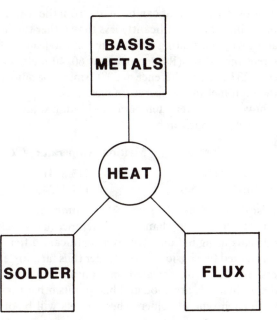

Figure 6.1 Requirements for a good solder connection.

The melting point of a particular solder depends on its composition. Consider common tin/lead solder. Most people take this solder for granted, since it is the most common type available and has so many uses. The most widely used tin/lead solder (60/40) has a melting point of 370°F (187.7°C). If you use a tin/lead solder with a 50/50 composition (50 percent tin and 50 percent lead) you increase the melting point to 417°F (213.8°C). If you go one step further and use a 40/60 composition of tin/lead solder, the melting point goes to 460°F (237.7°C). A small variation in content of the solder can cause significant changes in the temperature at which it melts. This, of course, is also true of solders with different compositions of indium in them.

Indium, as discussed in Chapter 3, is a soft metal that has many properties beneficial for microwave applications. One property of indium that makes it useful for solder application is its low melting point. Consider, for example, the melting point of indium as compared to tin and lead. This comparison is shown below:

Element	Melting Point
Indium (In)	156°C (312.8°F)
Tin (Sn)	231.9°C (449.4°F)
Lead (Pb)	327.4°C (621.3°F)

There are two observations that can be made from the listing above: (1) The melting point of indium is significantly less than either tin or lead, and (2) it now is apparent why the melting temperature of tin/lead solder increased as the lead content increased. (Remember that 60/40 melted at 187.7°C and 40/60 melted at 237.7°C, a difference of 50°C simply because of the increased content of the high-melting-temperature metal: lead.)

To realize how much lower a temperature indium solder will take to melt, consider the direct comparison below:

Solder	Melting Temperature (°C/°F)
Sn/Pb (50/50)	213.8/417
Sn/In (50/50)	117.2/243

This comparison shows that the same 50/50 combination of tin and either lead or indium results in more than 96°C difference in the melting point. The lower, of course, is using indium. This is a significant difference.

We have presented two basic types of solder thus far: tin/lead and indium-based solders. We have concentrated only on melting temperatures to this point, but there obviously are good and bad points other than temperatures, associated with each type of solder. These points will be our next topic of discussion.

Tin/lead solder, as previously mentioned, is available in a variety of compositions. Most texts on solder list no less than six combinations of tin and lead solder:

- 63/37
- 70/30
- 60/40 (the most common)
- 50/50
- 40/60
- 20/80

(The numbers indicating the percentage of each metal, tin and lead.)

Melting temperatures for these compositions are as follows:

63/37	361°F (182°C)
70/30	367°F (186°C)
60/40	370°F (187°C)
50/50	417°F (214°C)
40/60	460°F (238°C)
20/80	531°F (277°C)

The type of tin/lead solder that you use will depend on your application as far as such things as metallurgical make up of the devices to be connected and the temperature limits of these devices. It would not be wise, for example, to use 40/60 tin/lead solder on a transistor that had a maximum temperature range of 200°C, since it takes a 238°C temperature to melt the solder. The temperature range of the devices you are soldering must always be considered.

Similarly, as mentioned above, you should be aware of the metallurgical make up of the devices you are soldering. Tin/lead solder should not be used to solder to gold or silver plating because the tin tends to scavenge gold and silver. You will recall from Chapter 2 that gold could not be put directly on the alumina substrates because of a scavenging process; an adhesive layer was needed to make the complete substrate. This same "leaching" process will occur with tin/lead solder on gold or silver plating unless the solder has small amounts of gold or silver to reduce this phenomena (62Sn/36Pb/2Ag, for example). A very brittle AuSn (gold/tin) intermetalic will be formed, which usually causes connection failure in extended temperature cycling. Therefore, pure tin/lead solder should be used only when the contacts on a component or the ground plane of a substrate has a tin plating or is of metallically compatible construction. Even though we have said that tin/lead/silver (62Sn/36Pb/2Ag) solder can be used on gold, it is a good idea to stay away from that process and use an indium-based solder instead, just to be safe.

Tin/lead solders are higher-temperature *soft* solders that find uses in microwaves to attach components (chip resistors and capacitors) and also to attach substrates to cases. When using tin/lead solder, however, remember to check the metals that are being soldered so as to prevent a brittle connection that could cause problems later on.

Indium solders are generally lower-temperature solders that are much more flexible than tin/lead. This flexibility is due to the softness of indium (described in Chapter 3), which generally is an advantage but càn cause problems by being too soft.

Typical solders with an indium base are tin/indium (SnIn, 50/50), lead/indium/silver (PbInAg, 15/80/5), and lead/indium (PbIn, 50/50). Although there are many more types of indium solder, these are the most common types used for microwave applications. It is interesting to note that Indium Corporation of America in Utica, New York, offers a variety of kits containing different types of indium solder. There is one offered that is called the *microcircuits kit*. This kit contains the following solders:

Indalloy No.	Composition	Temperature (°C/°F)
290	97 In, 3 Ag	143/290
2	80 In, 15 Pb, 5 Ag	149/300
4	100 In	157/313
204	70 In, 30 Pb	174/345

205	60 In, 40 Pb	185/365
7	50 In, 50 Pb	209/408
206	60 Pb, 40 In	225/437
10	75 Pb, 25 In	264/508
150	81 Pb, 19 In	280/536
164	92.5 Pb, 5 In, 2.5 Ag	300/572

One thing to notice about all of the above solders in this kit is that the only elements used are indium (In), lead (Pb), and silver (Ag). There is an obvious lack of tin in any of the solders because, as you recall, the tin reacts with gold to form AuSn metallics that are very brittle and cause connection failure. Since most microwave components use a gold plating for cases and conductor coating, you do not want tin in the area to produce this condition.

As an example of actual applications of indium solder you could use the Indalloy 7 to put a substrate onto a carrier plate or case bottom and use an Indalloy 2 to attach the components to the substrate. This would put less thermal stress on the component (143°C) and use the higher-temperature solder (209°C) for the larger ground plane area. In addition, the higher-temperature solder would be used so that the ground plane would not come unsoldered when heat was applied to attach the components.

Before leaving the topic of indium solders we should mention that just as tin and gold should not come in contact with each other, indium and copper should also be kept apart. If contact is make, the copper will diffuse into the indium and cause an unreliable connection. This interface should therefore be avoided.

So we have looked at the first block of Figure 6.1: solder. The above discussions have covered two of the most prominent types of solders used in microwaves: tin/lead and indium-based solders. Table 6.1 is a summation of some of the typical solders used for microwaves. They are listed by composition and in increasing order of melting temperature.

Table 6.1
Microwave Solders

Composition	Melting Point (°C/°F)
50 Sn, 50 in	125/257
15 Pb, 80 In, 5 Ag	149/300
100 in	157/315
63 Sn, 37 Pb	182/361
70 Sn, 30 Pb	186/367
60 Sn, 40 Pb	187/370

50 Pb, 50 in	209/408
50 Sn, 50 Pb	214/417
40 Sn, 60 Pb	238/460
75 Pb, 25 In	264/508

Sn = Tin
Pb = Lead
In = Indium
Ag = Silver

With a topic of solders completed, the next block from Figure 6.1 is that entitled *flux*. The main job of solder flux is to remove impurities from the metallic surface to be soldered so that the soldering process can take place as efficiently as possible. The impurities it removes are usually oxides that have formed on the metal surface. A very common oxide is one that is familiar to all of us: iron oxide, or rust. The flux ensures that oxides or other impurities do not interfere with the soldering process. To be an effective "cleaning" agent, a flux should accomplish the following tasks:

- Provide a liquid cover over the material to be soldered and shut out any air up to the solder melting temperature. (Any air will increase the creation of oxides on the metal.)
- Dissolve oxides on the metal surface and carry unwanted material away (basically, make the solder area free of any impurities that will hinder connection of the metals to be joined). This is part of what is called a *wetting* process: the process that forms a uniform, smooth, and adherent film to allow soldering to a base material.
- Be easily displaced from the metal when the solder reaches its fluid state. (Once the area is cleaned by the flux, the flux must move out of the way to allow the soldering process to take place.)
- Be easily removed after the soldering process is complete. (This provides a clean connection.)

There are three types of fluxes used in soldering: corrosive (inorganic), intermediate (organic), and noncorrosive (rosin):

- *Corrosive (inorganic) flux*, designated IA, is not used for electronic assemblies. It is used for metal alloys or stainless steels that are difficult to solder. This type of flux is very corrosive and can damage other components that are in the area you are soldering if great care is not taken. You will not see this type of flux used in microwave or any other electronic applications.

- *Intermediate (organic) flux*, designated OA, is divided into three sub-categories: organic acids, organic halogens, and amines and amides. All three subgroups are corrosive (the organic halogens are more corrosive than the rest) and are very temperature-sensitive. Most produce condensed fumes that must be carefully disposed of to ensure operator safety. These fluxes, like the corrosive fluxes, are not used for microwave or electronic applications.
- *Noncorrosive* (rosin) *flux* is the most frequently used flux for microwave and electronic application. The rosin flux is divided into three subgroups also: water-white rosin, designated R; mildly activated rosin, designated RMA; and activated rosin, designated RA.

The safest and mildest flux is the water-white rosin (R) type. It is used on very clean surfaces where gold and silver (highly solderable metals) are utilized. Any residue left after soldering creates no corrosion problem. It can, however, be removed with a solvent such as trichloroethane. (1,1,1 trichloroethane is the minimum concentration of solvent used. Higher-purity solvents are also available for use.)

The mildly activated rosin flux (RMA) has a shorter wetting time than R-fluxes and is used, once again, on gold, silver, and also copper. These fluxes are noncorrosive and nonconductive and are removed from the finished circuit only in very critical applications such as aerospace equipment. When it is necessary to remove the flux, a combination of alcohol mixed with 1,1,1 trichloroethane can be used.

Activated rosin fluxes (RA) are used on metals such as nickel and cadmium. They are the strongest and most active of the rosin fluxes and are sometimes used to speed up the soldering time of such metals as gold. Residue from these fluxes should always be removed with a solvent such as isopropanol (alcohol) and 1,1,1 trichloroethane.

It should be clear at this point that the choice of flux for a solder connection is at least as important as the choice of solder. In many cases it is more important, since the worng flux will not remove the oxides or impurities, may not flow out of the way of the solder, and may leave a residue that will be harmful to either the circuit being soldered or an adjoining circuit. Great care must be taken when choosing the flux to use.

The final block shown in Figure 6.1 is called *basis metals*. These, very simply, are the metals being joined. We have previously mentioned metallic combinations that will result in an improper solder connection. Tin and gold or indium and copper are examples of incompatible metals. You should be sure of the metallic makeup of the items to be joined prior to attempting to join them. Many times manufacturers of chip components (resistors and capacitors) will list recommended methods of attachment and will give solder types you should use. Take note of these recommendations, and follow them as closely as possible.

The topic of *heat* in the soldering process has been covered over and over again. You should make this one of your primary considerations when considering a solder connection. Some tables have been presented in this chapter, and solder vendors, of both tin/lead and indium, with gladly provide literature with additional tables of temperatures.

We have covered all aspects of the requirements of a good solder joint: solder, flux, basis metals, and heat. Let us now see how these work in an actual application. The best application is one where some problems can occur if care is not taken in choosing the right solders, flux, metal, or heat. That application is *soldering to gold.*

As previously mentioned, soldering to gold is an operation that requires care. If there is a tin content to the solder, a brittle connection results. If the gold layer is so thin that it is removed by the soldering process and there is copper below it when you are using an indium solder, there will also be problems. This can be a difficult connection to make, but if careful consideration is given to all aspects of the connection to be made, it can be a fairly routine matter.

Let us use a specific example. Suppose we have to solder a substrate consisting of chrome/copper/gold metallization to a gold-plated kovar carrier plate. The metallization has the following dimensions:

Chrome	200 to 300 Angstroms (Å)
Copper	250 ± 50 microinches
Gold	50 microinches (min)

We also have requirements that say we have to remain below 220°C temperature. Table 6.2 shows six common solders that are used in many electronic applications.

Table 6.2
Common Solder Characteristics

Composition	Temp (°C/°F)	Coefficient of Exp. (PPM/°C)	Flux
50 In, 50 Sn	125/257	20	RMA/R
80 In, 15 Pb, 5 Ag	142/290	10	RMA/R
100 In	157/313	29	RMA/R
63 Sn, 37 Pb	183/361	25	RMA/R
60 Sn, 40 Pb	188/370	27	RMA/R
50 In, 50 Pb	209/408	27	RMA/R

As we start our search for the proper solder we can eliminate three solders immediately because of their tin (Sn) content: numbers one (50 In/50 Sn), four (63 Sn/37 Pb), and five (60 Sn/40 Pb). Our previous discussion explained how tin and gold will form a very brittle and unreliable connection.

Number six (50 In/50 Pb) would probably be eliminated because its melting temperature (209°C) is approaching the maximum limit imposed by our initial requirements. If you can guarantee that the 220°C figure would not be approached, this solder could be used. Usually, however, you should allow some sort of safety factor when using solders. This solder should be eliminated for this application.

Solder number three (100 In) would probably also be eliminated since pure indium solder is very soft and should not be used for attaching substrates to a carrier. There are times when a small chip component (capacitor or resistor) can be attached with 100 In, but a substrate has too large an area for pure indium solder to be used.

The one remaining solder, number two (80In/15Pb/5Ag), fulfills all of the necessary requirements to solder over gold metallized substrate:

- It contains no tin (Sn).
- It has a small content of silver (Ag) to aid in a good gold solder connection.
- It melts at a temperature substantially below the maximum 220°C temperature (142°C). This will not dissolve the gold like higher temperature solder will.

(Note that solder also has a coefficient of expansion of from one-half to one-third that of the other solders. This may be a benefit or a liability, depending on application. Be sure to check this out before using any solder for any application that has to operate over a large temperature range.)

This is one example of solder used to solder to gold. There are some conflicting views on exactly how to accomplish this task. Most papers published on this topic favor indium solder and deal with three concepts concerning soldering to gold: *scavenging, wetting,* and *aging.* These terms are very important when choosing a solder for gold applications.

There are also some papers that say that tin/lead can be used to solder to gold if the gold content of the solder connection is kept very low. Most testing shows that this is a critical balance that few people can achieve. For this reason a good rule to follow is to use solder with *no tin content* when soldering to gold.

Let us backtrack a bit here to look at the three concepts listed earlier: scavenging, wetting, and aging. The first two have been mentioned previously but will be covered in more detail here to show their relation to gold soldering.

Scavenging is the dissolving of metallization by the liquid solder that makes the soldered pieces useless. Tin/lead solders cause significant scavenging on gold. Numerous tests have been run that compare the gold/lead/tin interface and the gold/lead/indium interface. An overwhelming number of these tests show that there is considerable scavenging with the gold/lead/tin combination and little or no scavenging with the gold/lead/indium combination: an excellent reason to stay away from tin/lead solder on gold.

Wetting is the characteristic of liquid solder in contact with a metal part that causes flow and spreading of the liquid until an equilibrium point is reached. Basically, it is the ability of a solder to flow properly to provide a good connection over the entire contact area. This is important in the soldering of gold, since a solder that wets very poorly will take more time (and heat) to work properly. This will tend to dissolve the gold, and the final connection will not be the gold but some other base metal. A good wetting solder is therefore the desired type to use.

The aging process is the change in a solder connection that occurs because of a thermodynamic interaction between the connection and the environment it is exposed to. This interaction causes metallic compounds to grow at the connection. Such intermetallic compounds an $AuIn_2$ and Au_9In_4 form and can impair the mechanical bond of the solder connection. For this reason indium/lead solders should not be used when:

- Solder is less than 15 mm thick or
- Gold film is thicker than 10 mm.

In this case the 80 In/15Pb/5 Ag solder we previously chose for our example should be used. The 5 percent Ag (silver) will cut down the development of these intermetallic compounds.

All soldering is not done with a hot iron. Many times substrates and components are attached by means of a *reflow* process. In this process preforms of solder are placed between the device to be soldered (substrate or component), and the ground plane or substrate and the entire assembly is heated. The solder then flows throughout the open space between the units. This process can be controlled very closely and is a good method for covering larger areas with solder.

Another method that can be used is *vapor-phase soldering*. This process is also a reflow process in that it melts a solder that has been previously applied. In this process the pieces to be soldered are "immersed" in an atmosphere of saturated vapor from a boiling liquid. The vapor from the boiling fluid completely envelops the pieces to be soldered and begins to condense, giving up its latent heat of vaporization. This heat rapidly and uniformly raises the temperature of the pieces to be soldered to the boiling point of the liquid and thus melts the solder. Figure 6.2 shows a basic vapor-phase system. This type of system is good for some application where a bit of soldering is needed.

At this point you should realize that both tin/lead and indium solders have very definite and acceptable places in microwaves for a variety of solder applications. The trick is to find the right solder for your application that will provide a solid mechanical and electrical bond that will be reliable for many years of operation.

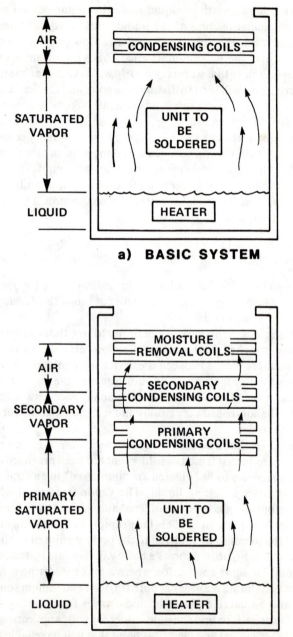

a) **BASIC SYSTEM**

b) **SYSTEM WITH SECONDARY VAPOR BLANKET**

Figure 6.2 Vapor phase soldering.

6.3 EPOXY

When solder is used, you are effecting a metallurgical type of bond; that is, two metals are actually being joined together. When epoxy is used you are developing an adhesive type of bond. This is the all-purpose glue for electronics and microwaves.

Epoxies are probably the most difficult types of substances for uninformed individuals to understand. Generally, when something is heated to a temperature of 100°C or higher it becomes a thin liquid and then hardens as it cools. Contrary to this concept, the epoxies actually harden (or *cure*) at these elevated temperatures and remain hard when the circuit is cooled. Many a program manager or components person has argued that a component or substrate must float when the epoxy is cured and cause inconsistent connections only to find out that the connection has actually gotten better as the epoxy cured. This is the first concept that must be grasped in order to fully understand epoxies and their uses.

The first and most obvious question to be asked is, What is epoxy? Epoxy is defined as a thermosetting material used for adhesive purposes. This definition is adjusted to fit the applications used in microwaves, since epoxies are also used for potting. We are, however, only interested in them as adhesives in this text. Epoxy comes in two types: one-component (part) and two-component. The one-component type is already mixed, while the two-part requires the mixing of the resin and hardener to form the final epoxy mixture. Both the resin and hardener of the two-part epoxy contain either silver or gold particles that must be distributed through the paste to ensure good electrical conductivity.

To aid in understanding this seemingly unorthodox type of material, we will present frequently used terminology for epoxy and define these terms in understandable language. We will then show typical epoxies that are used in microwaves to help you make the proper decision as to which one may be best for your particular application. Terms to be covered are *cure time, shelf life, pot life*, and *trixotropic.*

Cure time, very basically, is the time it takes for the epoxy to form a good electrical and mechanical connection. This time is determined by the composition of the epoxy and the temperature used for the curing process. Curing something usually involves a healing process, and you could relate this same type of process to epoxies. It is, in a sense, a healing process that gets the epoxy from a silver or gold paste material to the hard and solid electrical connection required for proper microwave circuit operations. You can think of it as being similar to concrete setting up from its flowing form into a solid form. Curing times can run from 10 minutes to two hours, depending on the makeup of the epoxy (one- or two-part) and the curing temperature used.

(These can range from a low of 80°C for some two-component epoxies up to 260°C for some single-component epoxies.) Typical cure times for two-part epoxies are:

- 10 minutes at 150°C,
- 20 minutes at 120°C,
- One hour at 100°C, and
- One and one-half hours at 80°C

and for one-part epoxies:

- One hour at 120°C,
- One-half hour at 150°C,
- 50 minutes at 177°C, and
- 15 minutes at 260°C.

Be sure to check the cure time of the epoxy you intend to use. It is important that the cure temperature does not exceed the maximum temperature of the components you are going to epoxy (resistors, capacitors, transistors, etc.). You also should check cure temperatures if you are doing two separate operations: that is, epoxying a substrate to a base plate and then epoxying components to a substrate (or vice versa). The first epoxy should have a higher curing temperature than the second one so that the initial connection will not be disturbed when the second curing process is executed.

The second term to be defined is *shelf life*. This is a length of time between shipment of the epoxy by the manufacturer and a date where you will no longer obtain the proper bonding and conductivity of the epoxy. Typical shelf life for one-part (component) epoxy is about six months when refrigerated and from one to two years for two-component epoxy. One difference between one-component and two-component epoxy is that the shelf life applies at room temperature (25°C) rather than requiring refrigeration to maintain its life. One important point to keep in mind is that the closer the epoxy gets to this shelf-life date the more questionable its operation will be. Always try to use "fresh" epoxy if at all possible.

Pot life is very similar to shelf life, except that the clock begins when you remove the epoxy from the original container as opposed to when the epoxy is shipped. Pot life is the length of time that the epoxy will be "good" after it is taken from its original shipping container.

Once again there is a drastic difference between one- and two-component epoxies when it comes to pot life. This time, however, the difference is reversed from that of shelf life. Pot life for one-component epoxy ranges from one week to three months, and for most epoxies it ranges from one to two weeks. The pot life for two-component epoxies, however, ranges from two to four *days*. If you are going to use two-component epoxy, you should mix only enough for the job at hand so that it will not be wasted if the pot life is exceeded. Even if you are using one-component epoxy, it is a good idea to remove only what you think you will need for the job. This will keep your assembly area from being cluttered with small batches of epoxy that probably should have been thrown out days earlier.

The final term to be defined is *trixotropic*. This term often appears an epoxy data sheet to describe its consistency. It is a term not usually found in any dictionary. The easiest way to think of the term *trixotropic* is to consider that it is a fluid that thins the epoxy paste so that it can be easily applied. Many thick film handbooks speak of screening resistive paste, a conductive paste on a conductive paste on a substrate, and say that a moderate degree of trixotropy is used to provide good definition of the screened area. They generally continue with a statement that highly trixotropic pastes tend to leave an imprint of the screen mesh; that is, they have thinned the paste too much so that it has more of a watery texture than a smooth, even paste. You can think of the term *trixotropic* as referring to a process for thinning epoxy so that it is easy to handle.

With the basic definitions used in epoxies presented, it is now time to investigate specific epoxies. As stated previously, conductive epoxy can be either a one-component type or may consist of two components. The one-component type is premixed and simply needs to be applied to the surfaces to be joined. The two-component type must be mixed in a prescribed manner and volume in order to result in the proper adhesive and conductive properties.

You may question the need for two types of epoxy. It would seem that the easiest type would be the one-component type where no additional mixing was necessary. In many cases this is true. There are, however, some disadvantages to the one-component epoxy that are eliminated by the two-component type. To distinguish between epoxies we will list advantages and disadvantages of each:

One-Component
- Short shelf life (six months);
- Needs refrigeration;
- Long pot life (two weeks to three months);
- Needs no parts mixing;
- Shorter curing times;
- Higher curing temperatures.

Two-Component
- Long shelf life (one to two years);
- Needs no refrigeration;
- Short pot life (two to four days);
- Parts must be mixed (resin and hardener);
- Longer curing time;
- Lower curing temperature.

Just as everything else we have covered in this text thus far has trade-offs, so also are there trade-offs to be made when choosing an epoxy. You should evaluate the pros and cons of each epoxy and choose the proper one for your application.

To understand what epoxies are commercially available, we will present one-and two-component epoxies from a variety of manufacturers and list terms a data sheet would show. Table 6.3 shows epoxies from four vendors —Epoxy Technology, Ablestik Laboratories, Emerson and Cuming, and Amicon (Polymer Product Division) — and lists a wide variety of epoxies with various values of pot life, shelf life, and curing times. One term that is not in this table is *volume resistivity*, which is a measure of the conductance of the epoxy. This value ranges from 0.0001 to 0.0003 ohm-cm for epoxy technology H20E and H20S; 0.0001 to 0.0005 ohm-cm for H40; 0.0005 to 0.0009 for H81; 0.0001 ohm-cm for Ablestik 58-1; 0.00004 ohm-cm for 36-2; less than 0.002 ohm-cm for Emerson and Cuming 58C; less than 0.0005 ohm-cm for Amicon CT 4042-5; and 0.001 ohm-cm for C-860-XCC. Once again, note the wide range.

Another term that should also be considered when choosing an epoxy is *lap shear strength*. This describes the strength of the connection regarding shear after curing. Values for the epoxies shown are:

Epo-TeK	H20E	1500 psi
Epo-TeK	H40	2000 psi
Epo-TeK	H81	2000 psi
Epo-TeK	H20S	1500 psi
Ablestik	58-1	2400 psi
Ablestik	36-2	1800 psi
E and C	58-C	1200 psi
Amicon	CT 404s-5	Not available
Amicon	C-860-3Xcc	1400 psi

A final term that should be covered in case the application is for space equipment is *outgassing*. This phenomena is created by a vacuum environment and causes a material to release vapors or gas when subjected to a vacuum condition that hampers circuit operations. The outgassing should be kept to a minimum in a good epoxy.

These are the terms used to describe conductive epoxy adhesives for microwave applications. As previously mentioned, you should evaluate each material as a separate entity, weigh the pros and cons, and make a decision as to which is best for your application.

With the terminology presented and defined, the next question may be, Where do you use epoxy? There are two application areas for epoxy in microwaves: (1) component attachment and (2) substrate attachment. Both of

Table 6.3
Common Epoxies

Manufacturer	Epoxy	Number of Components	Pot Life	Curing	Shelf Life	Consistency
Epoxy technology	H20E	2	4 days	5 min @ 150°C 90 min @ 80°C	1 year	Trixotropic paste (silver)
Epoxy technology	H40	1	—	30 min @ 150°C 1 hour @ 120°C	0.5 year	Trixotropic paste (gold)
Epoxy technology	H81	2	2 days	5 min @ 150°C 90 min @ 80°C	2 years	Trixotropic paste (gold)
Epoxy technology	H20S	2	4 days	5 min @ 150°C 90 min @ 80°C	1 year	Smooth-flowing paste (silver)
Ablestik Labs	58-1	1	2 weeks	30 min @ 150°C 60 min @ 125°C	6 months	Smooth soft paste (gold) needs refrigeration
Ablestik Labs	36-2	1	1 week	30 min @ 150°C 60 min @ 125°C	6 months	Smooth soft paste (silver) needs refrigeration
Emerson and Cuming	58-C	1	—	15 min @ 260°C 50 min @ 177°C	3 months	Silver paste
Amicon	CT4042-5	2	4 days	60 min @ 100°C 120 min @ 80°C	—	Silver paste
Amicon	C-860-3XCC	1	—	60 min @ 200°C	3 months @ 25°C	Smooth silver paste needs refrigeration

Note: Epoxy technology also has a flexible epoxy (H20F) that is used for flexible circuitry or for chip bonding where bonding temperatures are in excess of 250°C. Pot life, shelf life, curing time, and temperature are all comparable to other H20 epoxies.

these areas will be covered in detail later in this chapter. For now we can say that with the proper conditions set up and precautions observed, you may use epoxy very effectively for attaching microwave components (capacitors, resistors, etc.) to substrates and substrates to carrier plates or cases.

Now that we have become familiar with the terminology and basic applications, our next step is to learn how to use the epoxy. The first item to consider when planning to use epoxy is the preapplication cleaning. As in soldering, cleaning is important when making a microwave connection with epoxy. Of prime importance is the substance you use to do this cleaning. Absolutely *do not use alcohol* or alcohol products. A freon bath from about 23°C to 36°C for less than five minutes will do a good cleaning job, as will a brief immersion in trichlorethylene followed by a freon bath.

Chemicals such as toluene or xylene should *not* be used. Toluene has a high degree of toxicity, and xylene has cleaning properties that can be likened to cleaning with kerosene.

Other chemicals or solvents can be used prior to epoxy application. If you have a chemical you would like to use but do not know if it is suitable for use, call and discuss it with the epoxy manufacturer . They usually will discuss methods with you and make recommendations that will reduce, or eliminate, your problems and give you a well-behaved microwave system as an end product.

The next area of importance when working with epoxy is the *mixing*. (This section does not apply if you use one-component epoxies; they are already mixed for you.) When using two-component epoxies there is a definite way to combine the individual components (resin with silver or gold powder and hardener with silver or gold powder). Instructions for these epoxies say that they are mixed 1:1 by volume or weight; that is, equal quantities of resin and hardener. There is also an instruction below the mixing ratio on most data sheets that says "Mix the contents of part A (resin) and part B (hardener) thoroughly before mixing the two together."

This simple one-line statement is vital to achieving a reliable and lasting connection with two-component epoxy. If the two individual components (resin and hardener) are not *thoroughly* mixed *before* the two are mixed together, there will be substantial quantities of silver or gold powder that are not mixed throughout the paste. This causes areas of high resistance in some areas (where there is a low concentration of silver or gold powder) and areas of low resistance (high concentration of powder) in others. What we are after, ideally, is a material that has a high degree of consistency throughout the whole area. This can be accomplished only if each individual part is mixed thoroughly prior to mixing the two components together. Whenever you have to use two-component epoxy or have someone else apply it for you, be sure that it is mixed properly.

With the epoxy mixed, the next step is to apply it to the surfaces that are to be joined. Epoxy has been applied in a variety of ways over the years, with everything from screens to needles to toothpicks to the ultimate automatic machine. The screen and the needle are the most widely used types of applicator. A 200-mesh screen is usually used with approximately a 0.002 in layer of epoxy deposital. This would be the method used when you are epoxying down a substrate to a carrier plate or case. It is much more feasible when a large area is being epoxied.

A fine needle should be used to attach components with epoxy. This will dispense a precise dot of epoxy in the area of connection; this dot should usually not exceed 0.005 in. in diameter. The component is then placed on top of the epoxy dot and pressed into place. Care should be taken not to press too hard, as the epoxy can flow out from under the leads and possibly short a component to itself or to ground. A gentle pressing is sufficient.

To this point we have spoken of joining substrates to carrier plates, substrates to base plates, and components to substrates without too much consideration as to what material being joined together or what effect a certain material may have on the epoxy being used. Just as in the solders there are some metals that are just not compatible; for example, silver on aluminum, tin, or lead surfaces. You should *not* use silver epoxy on any of these surfaces (that is, on bare aluminum or a tin- or lead-plated surface); it results in no real bond between the surfaces and a brittle connection. This, in turn, results in a high-resistance connection that eventually cracks and separates (especially under environmental conditions). Gold epoxy also should not be used on bare aluminum for the same reasons. You should make a habit of using only screws to attach any substrate to bare aluminum.

Some people try to use an iridite (alodine) process that puts a coating over the bare aluminum. This process should not be used where high-reliability applications are involved or where a ground plane requiring high conductivity (which is always in microwaves) is needed. Whenever aluminum is involved, have it gold- or copper-plated to avoid immediate and long-term problems. This may sound expensive, but many times you can plate the base plate only and apply an iridite to the side walls of the case and have a circuit that works very well. Other times, however, a completely plated case is a necessity. This desision will depend, once again, on your application.

To sum up our discussions on epoxies we will list the qualities that make an epoxy suitable and acceptable for microwave applications:

- Low-volume resistivity (0.0001 to 0.0003 ohm-cm);
- Lap shear strength between 1,000 and 2,000 psi;
- Flowing properties for screening ease;
- Pot-life stability (uniformity in the batch);

- Good shelf life (at least six months);
- Minimum outgassing;
- Minimum bleeding (or spreading) of the resin;
- Curing temperature compatible with the substrate or component being used.

With these requirements met, you will have a very acceptable adhesive for microwave applications.

One statement appears on all epoxy data sheets that has not yet been mentioned: that is, a *caution* statement that reads something like this: "This product may cause skin irritation to sensitive personnel. If contact with skin occurs, wash the affected area immediately with soap and water." Pay close attention to any cautions printed on the data sheets, and observe them to the letter. Careful attention in the beginning will pay off later when accidents or painful experiences are avoided.

6.4 BONDS

This chapter first discussed soldering and classified it as a metallurgical bonding process and next covered epoxies and called this adhesive bonding. We will now go back to a metallurgical type of bonding and discuss *thermocompression, ultrasonic*, and *thermosonic bonding techniques*. Each of these methods is used for attaching microwave components to substrates. These are specialized types of bonds because they are not used to attach large areas such as substrates to carriers, as are soldering and epoxy methods; rather, these are wire bonding methods. They are designed to attach one area to another (usually from a chip to a connecting pad or a circuit). The bond is formed by placing a wire (or strap in some cases) on the appropriate metal pad and having a bonding tool with heat and pressure placed on this wire. The heat and pressure actually deform the wire (or strap) material and force it to be bonded to the metal pad. This bond comes about by the interaction of atomic forces between the wire and metal. The actual metal bonding process consists of three variables:

- *Pressure* (*or force*): To promote a plastic flow and close contact between metals;
- *Elevated temperture* (*heat*): To promote contaminate dispersal while lowering flow stress and improving the diffusion process;
- *Time* : To promote solid-state diffusion in the actual bond zone.

This process should not be confused with a commonly known term, *eutectic bonding*. A eutectic bond also involves heat and pressure, just as described above, but this type of bond requires a third metal film to be placed between the wire and metal pad. This produces a diffusion process when heat and pressure are applied. The bonds we are describing above, and throughout this section, are brought about by interatomic forces.

In the wire bonding process the wire itself must be ductile (be able to be drawn thin without breaking) and must be capable of deforming to comply with the bonding tools used. The metals that are most widely used in this process are gold (Au) and aluminum (Al), with aluminum being preferred in many applications because of its more rigid structure once the bond wires have been put in place. Diameters of 0.7 mils (0.0007 in) to one or two mils (0.001 in to 0.002 in) are common wires used for bonding.

The metal that the wires are bonded to must have a composition that does not have a high degree of oxidation. Materials such as copper (Cu) or palladium (Pd) are not generally used, since they are very susceptible to oxidation. Materials such as platinum (Pt), gold (Au), or silver (Ag) are used extensively, however, since they have very little tendency to oxidize. You can see how an oxidation on the surface of the metal could obstruct the bonding process, since it would be like an impurity film on top of the surface that would be between the two metals that were attempting to join. This would be much like not cleaning a circuit board prior to soldering or applying epoxy. The bond may take place initially, but its integrity and reliability would be questionable. We will now cover the three types of bonding processes listed previously: (1) thermocompression, (2) ultrasonic, and (3) thermosonic.

6.4.1 Thermocompression Bonding

Thermocompression bonding is almost self-explanatory. The two parts of the name say just what is involved in the bonding process: *thermo* means a temperature, and *compression* relates to pressure. Thus, thermocompression is accomplished by the use of a combination of heat and pressure to form the interatomic structure needed for wire bonding. The heat is applied either to the substrate, the chip being bonded, or both, while pressure is applied to the bond area. In this way both the temperature and pressure can be very closely controlled and applied to the appropriate area.

There are three basic types of thermocompression bonds: the *wedge bond*, the *ball bond*, and the *stitch bond*. Each of these has specific areas where it should be used. The type you choose, once again, depends on particular application, substrate temperature, or pressure limitations, and the metals involved in the bonding area.

Wedge bonding is shown in Figure 6.3. You will note that both the wedge and the substrate that is being bonded to are heated. This process, combined with the pressure applied by the wedge tool, provides an excellent connection. The wedge tool usually is made of a very fine sapphire or silicon carbide.

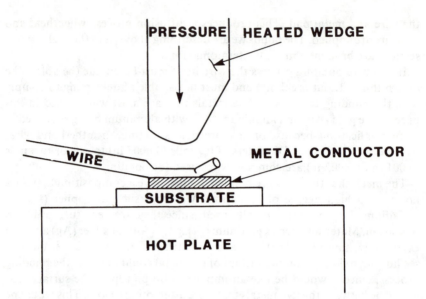

Figure 6.3 Wedge bonding.

Bonding occurs when the tool and substrate are at the proper temperature and the wire is in place. The wedge is then brought down for bonding, and the tool deforms the wire to form a highly reliable bond. After a specified time the tool is lifted, and the bond is complete. This type of bonding is rather like pressing a thin candle on a sheet of paper with a soldering iron: The combination of heat, pressure, and time have caused the wax to adhere to the paper. Similarly, the combination of heat, pressure, and time (all closely controlled) have bonded the wire to the metal conductor. Wedge bonding is used many times when very fine wire (less than 0.7 mils) is to be bonded.

The values used for pressure, temperature, and time will vary from device to device, and the manual for the particular bonder being used should be consulted. As a representative example, the following values are used to wedge bond to a transistor chip:

Temperature	300°C ± 10° for silicon bipolar,
	260°C ± 10° for GaAs FET's
Pressure	30 grams (40 grams, max)
Time	2 to 3 seconds

The values above are examples. Many times it takes trial and error methods to find the ideal bond for your application. This ideal bond usually is when the "footprint" produced by the bonding tool (the depression in the wire) is two to three times longer than the diameter of the wire. This is shown in Figure 6.4.

This configuration will result in a good solid bond that can be relied on to hold for many years of operation.

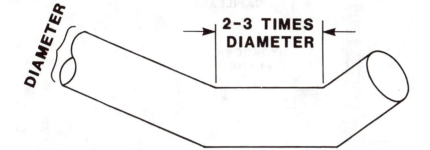

Figure 6.4 Ideal wedge bonding.

Ball bonding, also termed *nail head* bonding, uses thermocompression techniques like the wedge bond but does so through a capillary tube. This is a tube with a very small opening. This type of bonding, however, usually is not used for wires as small as 0.7 mils, since these fine wires do not flow smoothly through the capillary tubes. Generally wire of 1 mil or larger is used for ball bonding.

Figure 6.5 shows ball bonding. The bond is formed by first feeding a gold wire through the capillary and forming a ball on the end of the wire with a hydrogen flame. We specified gold wire because this is the only wire that will form the required ball. Aluminum wire cannot be used for ball bonding as it was for wedge bonding because aluminum will not form a ball when cut by the hydrogen flame. (The ball produced by the hydrogen cutting flame is shown in Figure 6.5a.)

After the ball is formed, the capillary is lowered until contact is made with the metallization. It is then lowered further where a predetermined force deforms the ball and causes the bond to be formed (Figure 6.5b) In this condition the substrate and/or the capillary are heated. Usually the heated probe tip (capillary) type bonder is preferred because the substrate does not need to be heated and there is no danger of destroying or damaging a component to be bonded. With this type of bonder the heat is generated by sending a pulse of current through the tip of the tool while the bond is being made. This means that the device is heated only in the immediate area of the bond being made. After a specified time, the capillary is raised from the substrate. This leaves the "nailhead" bond previously referred to (Figure 6.5c). The hydrogen flame once again heats the wire, cuts it, and forms a new ball for the next bond.

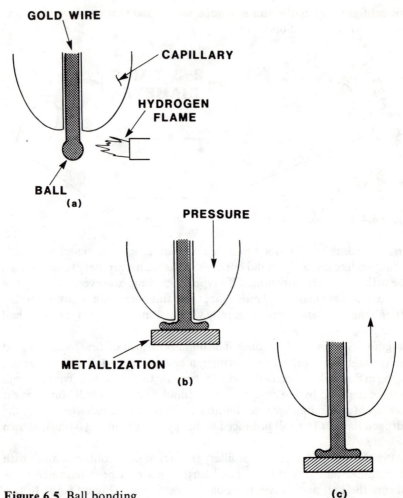

Figure 6.5 Ball bonding.

The third type of bond is the *stitch* bond, shown in Figure 6.6. The stitch bond is a combination of the wedge and ball bond. It is like the wedge bond because the wire is depressed to make contact with the surface to be bonded, and it is like the ball bond because a capillary is used. The stitch bond is begun by feeding the bond wire (either gold or aluminum) through the capillary tube and bending it at a 90° angle. This bend, and the absence of a ball, allows the use of either gold or aluminum wires, as previously mentioned. The initial bend is shown in Figure 6.6a.

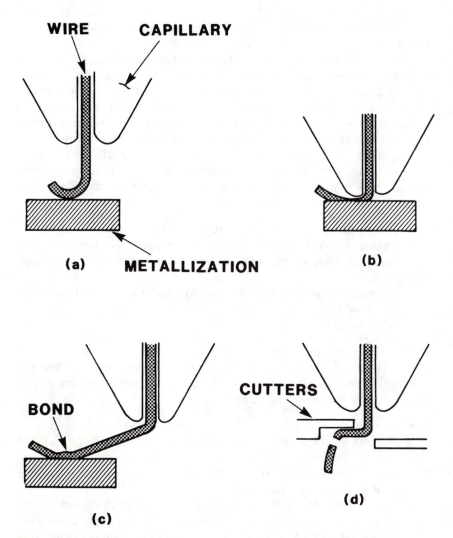

Figure 6.6 Stitch bonding.

The next step in a stitch bond is to lower the heated capillary and apply pressure to the bond wire to deform it and form the desired bond. (The substrate may also be heated, but you should be careful that the temperature used does not degrade the parameters of the circuit.) The deformation of the bond wire is shown in Figure 6.6b.

In Figure 6.6c the bond has been made, and the capillary is pulled back to begin making a second bond. In Figure 6.6d the bond wire is cut by metal cutters rather than the hydrogen flame bonding. This, once again, allows you to use either gold or aluminum wire for your bonds.

One very important point must be brought up at this juncture: The bonding tool must be moved *in the direction of the stitch* when going from the first bond (Figure 6.6b) to the second (Figure 6.6c). If it is not moved in this direction, a stress point is created directly at the bond that more than likely will cause a bond failure. It is vital, therefore, to position the substrate so that the first and second bonds are directly in line. This can be considered to be a disadvantage of the stitch bond in that it must be so critically positioned.

These are the bonding methods that are available when using thermocompression bonding. In review, the following requirements should be met to ensure a good and reliable bond:

- Choose the appropriate wire material (gold or aluminum).
- Be sure that sufficient deformation of the wire occurs when making the bond. The bond length should be from two to three times the diameter of the wire.
- Choose, and optimize, the time, pressure, and temperature to make the best bond for your application.
- Be sure that the bonding surface is clean.

If the above procedures are followed, you should have no trouble with your bonds. You should, however, be aware of some of the causes of poor thermocompression bonds. The three primary causes of poor bonds and a brief explanation of each are listed below:

- *Too much or too little heat or pressure*: These conditions can be detected by examination under a microscope. They will show up as a bond wire that has either very little deformation or excessive deformation. Remember the two-to-three-times-the-wire-diameter number quoted previously for an ideal bond.
- *Impure wire*: This will show up as a bond wire that will not deform properly. Even when increased pressure and heat are applied, it will not deform properly. This is because even the smallest amount of impurity within the wire will cause it to harden and will not allow it to deform to produce the proper bond.
- *Dirty or glassy metallization*: This will cause a properly deformed bond wire to either not bond at all or form a very weak bond. All oil, dirt, and oxides must be removed from the metallization before you attempt any bonding process.

One additional problem that can arise when bonding aluminum to gold is what is called the *purple plague*. This is a brittle alloy that forms when an aluminum/gold (Al-Au) bond is exposed to extended heat. This is why both the temperature and duration are so important when bonding aluminum and gold. This problem causes frequent bond failure because the gold is absorbed into the newly formed alloy and adhesion between the wire and the metallization is lost. Take great care when choosing a bonding temperature *and* duration.

6.4.2 Ultrasonic Bonding

Ultrasonic bonding is a *cold* bonding process: That is, there is no heat associated with the joining of two metals other than that set up by a scrubbing action of the bonder. This type of bonding has many advantages where delicate circuitry should not be exposed to elevated temperatures that can be associated with other bonding techniques.

Figure 6.7 shows an ultrasonic bonding setup. You can see that it looks much like all of the other bonding methods that use a capillary tube but has two exceptions. There is an ultrasonic head attached to the capillary tube, and there is absolutely no reference to temperature on either the substrate or the capillary tube. This is because the process uses a scrubbing action and pressure to join the two metals and no heat is needed to deform the wire to form the bond.

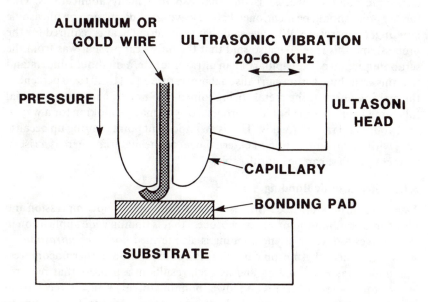

Figure 6.7 Ultrasonic bonding.

Ultrasonic bonders can use both aluminum (Al) and gold (Au) wires as well as silver (Ag) and copper (Cu). Both ball and stitch bonds can be made with the same restriction placed on ball bonding as in thermocompression bonding: That is, aluminum wire cannot be used because of its inability to form the required ball when cut by the hydrogen flame.

The advantages of ultrasonic bonding are many and make this process one that should be considered for production application:

- Requires no external heating;
- Has rapid bond rates;
- Is low cost;
- Allows easy replacement of chips without damage.

As with everything, the ultrasonic bonding process also has disadvantages:

- Settings on the bonds are very critical;
- The ultrasonic oscillator will drift if not left on to stabilize;
- Rough and uneven bonds can occur;
- Alignment is more difficult with this process.

If the disadvantages listed above can be overcome or minimized, the ultrasonic bonding process is one that can find many applications. One further point should be mentioned before leaving the discussion on ultrasonic bonding: temperature. We have said that no external heat is required for the process, and this true. We also said that the only heat present was from the scrubbing action of the capillary at an ultrasonic rate. You should understand that this scrubbing action can cause a temperature equal to 30 to 50 percent of the melting point of the metals being joined. Do not reach for the material following an ultrasonic bonding process, or you may be looking for a way to cool your hand rather quickly. This is an important point to bring up because even without an application of external heat to the material, there is a rise in temperature that can be considerable.

6.4.3 Thermosonic Bonding

Thermosonic bonding takes the best of the worlds of thermocompression and ultrasonic bonding and produces a process that is finding wide application in microwaves today. This type bonding is also termed *hot work ultrasonic.*

As mentioned, thermosonic bonding is a combination of thermocompression and ultrasonic bonding and as such results in a process that requires lower bonding temperatures and lower bonding forces. You will recall from the section on thermocompression bonding that typical values for bonding to a transistor chip were a temperature of 300°C for a bipolar device (260°C for

GaAs FET); pressure of 30 grams (40 grams max); and duration of two to three seconds. With thermosonic bonding these figures can fall into the range of 150°C temperature; a pressure of approximately 20 grams; and a duration still in the neighborhood of two seconds. We can get reduced temperatures and pressures because the thermocompression mode of the bonding process can operate at a lower temperature (150°C) with the ultrasonic scrubbing energy making up the difference for temperature and bonding pressure: That is, the bonding area is heated just high enough with the capillary to begin a bonding process. Then the ultrasonic movement of the capillary completes the process, resulting in a lower temperature and pressure.

Bonds produced with thermosonic bonding are the ball bond and the stitch bond. (Remember, once again, that aluminum wire cannot be used for ball bonding in the thermosonic process, either.) These bonds are fabricated exactly as described previously and will exhibit the same properties as discussed.

6.5 COMPONENT AND SUBSTRATE ATTACHMENT

With the various bonding methods now presented, it is time to see how these methods apply to actual microwave applications. The material presented here will be representative examples and recommendations for attaching active and passive components and substrates. For specific procedures you should consult the component or substrate manufacturer for directions for your particular application.

There are basically two methods of attaching a substrate to a carrier plate or case: solder and epoxy. Solder is usually applied by a reflow process so that a uniform coating can be achieved. Epoxy should be applied by a screening process so that this same uniformity can be achieved. This uniformity is of prime importance in the attachment of substrates, since the bond not only is holding the substrate in place but is providing a *ground plane* for the circuit. Be sure that a uniform flow of solder or thickness of epoxy is applied to attach a substrate. These techniques are valid for PTFE/glass, high-dielectric Teflon^R laminates, and alumina substrates. The one point to remember is one that was brought up before: Do not attach any substrate (solder or epoxy) to an unfinished aluminum plate or chassis. Contact will be inferior and the ground plane will not be continuous with this arrangement. Also, be aware of any intermetallic conflicts such as the gold/tin problems listed in previous sections. These can cause brittle and unreliable connections that may appear adequate when first made but will deteriorate rapidly with time and environmental conditions. Be sure to consider all possibilities when deciding which material to use to attach your substrates.

Passive components, such as chip capacitors and resistors, can be attached to a substrate in a variety of ways. The method used, of course, depends on the terminations (connection points) on each individual component. Chip resistors, for example are available with platinum/gold, silver over nickel, or solder terminations (the solder terminations are optional and must be specified separately). Both of the standard terminations for chip resistors can be attached with the same type of solder or epoxy:

- *Solder*: Reflow process using 60/40 or 63/37 tin/lead solder. If an iron is used you should not exceed 35 watts with a small chisel point.
- *Epoxy*: Silver-filled conductive epoxy is preferred with a curing temperature *below* 150°C.

Chip capacitors are somewhat more involved, since they are available with a variety of terminations. Typical combinations are:

- Platinum gold
- Gold over chromium
- Platinum silver
- Gold over nickel

Attachment for each of these terminations is as follows:

- *Platinum gold*: Solder with SN60, SN62, or Indalloy 7 (50 percent indium, 50 percent lead); eutectic bonding; or any standard gold- or silver-filled epoxy. (Note: Check the curing temperature of the epoxy you choose, and be sure it does not exceed that of the capacitor.)
- *Gold over chromium*: Thermocompression, ultrasonic, or thermosonic gold wire or aluminum silicon wire bonding; eutectic bonding; and standard gold- or silver-filled conductive epoxies with the same temperature restrictions for curing as mentioned above. Special solders can also be used that have a low tin content or no tin at all. (Recall the tin/gold reaction mentioned previously.)
- *Platinum silver*: Solders such as SN60, SN62, and Indalloy 7 (50 percent indium, 50 percent lead); standard gold- or silver-filled conductive epoxies.
- *Gold over nickel*: Thermocompression, ultrasonic, and thermosonic gold wire or aluminum silicon wire bonding; special solders that have low tin content or no tin; standard gold- or silver-filled epoxies.

A passive device that requires wire bonding for its operation is the Lange coupler. Figure 6.8 shows a drawing of the coupler and where the wire bonds are to be place. Figure 6.9 is an actual picture of the coupler and its bonds. The bonds used in this particular case are wedge bonds.

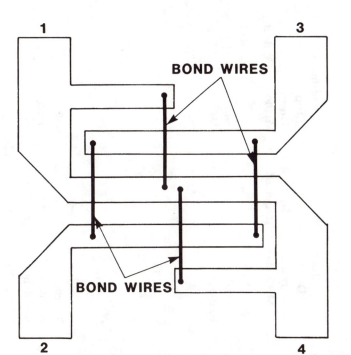

Figure 6.8 Lange coupler.

Courtesy of ITT

Figure 6.9 Lange coupler bonds.

Table 6.4

		Chips with or without Leads			Beam Leads		Stripline	
Type of Board	**Usable Die Down Methods**	**Best Method**	**Best all Around Method**	**Die Down Method**	**Tools Required**	**Strip Down Method**	**Top Contact**
Teflonr / Fiberglass	1. Soft solder	Hot gas bonder		Not advised to use		Soft solder (180° max.)	Parallel strap welding wire
	2. Conductive epoxy	Epoxy machine	Soft solder (180° max.)				
	3. Eutectic solder (usually too hot)	Hot gas bonder					
Ceramic	1. Eutectic solder	Hot gas bonder	Soft solder or eutectic solder (280° max.)	Thermal compression wedge bond or parallel gap weld	Beam lead bonding tool	Soft solder	
	2. Soft solder	Hot gas bonder					
	3. Epoxy	Epoxy machine					

Active devices require more care than passive devices when considering a method of attachment to a circuit. You need to be concerned not only with intermetallic interactions but also with the temperature used for attachment, since too high a temperature can damage sensitive devices. Such components as diodes and transistors must therefore be watched very carefully.

Table 6.4 shows attachment methods for microwave diodes on a variety of boards. The term *die down* refers to the putting of the chip into a circuit. *Strip down* is with the attaching terminals under the diode, and *top contact* is with the terminals on top of the device necessitating a strap connection.

It should be kept in mind that the methods shown in Table 6.4 are recommendations and are usually the best methods to use. You should, however, consult the individual data sheets or manufacturer if you have any questions as to which is the best method for your particular diode and application.

Transistor attachment is similar to that of microwave diodes. The major differences in transistor attachment, however, come about from your choice of device: silicon bipolar or GaAs FET. For example, consider the following typical method of die attachment for a silicon bipolar transistor first and a GaAs FET second (both are done in an inert atmosphere):

- *Silicon bipolar*
  1. Heat to 400°C ±10° the heater block.
  2. Place the circuit that will hold the transistor on the heater block, and allow it time to heat (usually five to 15 seconds will do).
  3. Place the chip (using tweezers) carefully on the circuit, and scrub back and forth until wetting occurs.
  4. Following wetting, do one circular scrub of the chip, and carefully remove the tweezers. A eutectic bond has now been formed.
  5. Remove the entire circuit from the heater block and allow it to cool naturally.

- *GaAs FET*
  1. Heat to 300°C ±10° the heater block.
  2. Place the circuit on the block and allow it to heat thoroughly (five to 15 seconds).
  3. Place a gold/tin (AuSn) preform on the circuit in the proper area.
  4. Using tweezers place the chip in the proper orientation on the circuit and scrub it back and forth until wetting occurs.
  5. When wetting occurs, make one circular scrub of the chip, and remove the tweezers. This is now a solder bond.
  6. Remove the circuit from the heater block, and allow it to air cool.

You now should be able to see the differences between silicon bipolar (a eutectic bond) and GaAs FET's (a solder bond) regarding bonding the transistor chip.

After the chip is in place, the next step is to get from the transistor to the external circuit around it. Wire bonds are used for this purpose (either a gold ball bond or thermocompression (wedge) bond). For higher frequencies the wedge bond is preferable because of its overall shorter length, which reduces electrical parasitics. The bonds are made as outlined in the previous sections, and specific settings and operations should be obtained from the individual bonder instruction manual. (Note: If you are required for one reason or another to use aluminum wire for bonding, be sure to keep close watch on the bonding duration to avoid creation of the purple plague, which we referred to previously. You will recall that this can take place at an aluminum/gold interface and degrade the bond greatly.)

6.6 CHAPTER SUMMARY

This chapter began by saying that bonding was probably the most necessary, and least understood, process used in microwave circuit fabrication. You should by now realize how necessary it is and also have a good understanding of what bonding is and how it is accomplished.

We have covered such important areas as solder, epoxy, and wire bonding in an effort to acquaint you with the different methods available for assembling microwave circuits. All of these methods were summed up by showing where and how they are used in attaching both substrates and components to microwave circuits.

At this point we can only caution you to treat each attachment in microwaves as an individual and unique case. Analyze what you are trying to bond, and take inventory of what materials and methods you have available to you, and you should have a circuit that will have reliable and dependable connections.

CHAPTER 7
MICROWAVE PACKAGING

7.1 INTRODUCTION

In most areas of electronics, the primary function of the package is to enclose a previously designed circuit. In microwaves, however, the package is an integral part of the overall circuit design and can mean the difference between a circuit that works and one that does not work.

To understand how important the package is in microwave applications, refer to Figure 7.1. The figure shows three methods used for microwave transmission: microstrip, stripline, and suspended substrates. In each case you see a dielectric (ϵ), a conductor, and a shaded area that is a ground plane. The last area mentioned (the ground plane) makes the difference in microwave packaging. You may have seen some or all of these configurations and referred to the shaded area as a ground plane (as it is), thinking of it as a piece of copper bonded to a laminate or as a block of aluminum used as a mount for your circuit. This block may also have seemed to be a convenient place to mount SMA connectors. This convenient shaded area, however, is much more than a connector or laminate support.

Refer once again to Figure 7.1, and notice differences in how the ground plane is oriented. The microstrip has a ground plane supporting the dielectric with air on top of the circuit; the stripline configuration has a ground plane both above and below the circuit; and the suspended substrate has a ground plane on the bottom of the dielectric just as in the microstrip case but with an air gap between the dielectric and the ground plane.

153

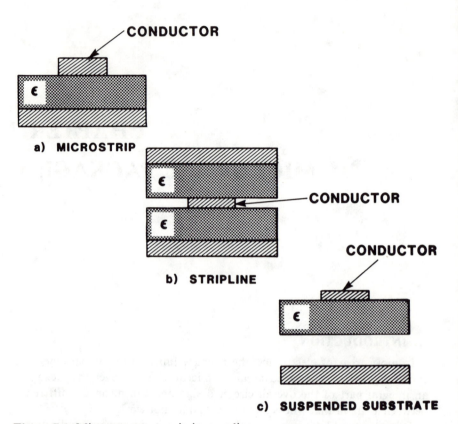

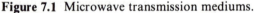

Figure 7.1 Microwave transmission mediums.

The transmission media shown in Figure 7.1 are representations of each configuration showing the orientation of individual parts. Let's put each of them in a package and show what each would look like under final packaged conditions. These are shown in Figure 7.2.

Figure 7.2a is packaged microstrip. Notice that the dielectric thickness is designated by *h* and the dielectric is placed directly on the case floor. The dimension *H* is the distance from the dielectric and the top of the case. This is a critical dimension and will be covered in detail later in this chapter.

Figure 7.2b is packaged stripline. This is the basic stripline package shown in Figure 7.1 with the two ground planes connected together to form a case. The *b* dimensions are those of the dielectric. You can see how this configuration forms a well-shielded circuit with dielectric and case completely surrounding the circuitry. Various methods of packaging stripline will be covered in this chapter.

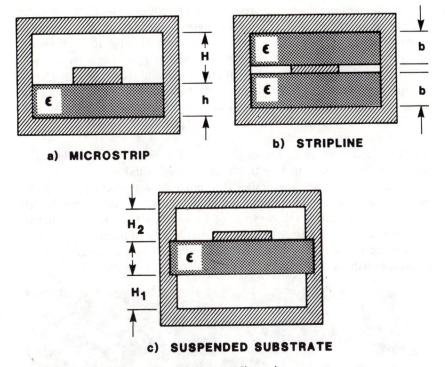

Figure 7.2 Microwave transmission mediums in cases.

Figure 7.2c is packaged suspended substrate. This is a method used in millimeter application. We will not cover this method of transmission medium in any great detail, since it is for higher frequencies and also resembles microstrip in many ways. Typical dimensions for suspended substrate are as follows:

Dielectric thickness	0.005 in
Dimension H_1	0.010 in
Dimension H_2	0.010 in
Dielectric width	0.046 in

The dimensions are relatively small and indicate higher frequency operation. Circuits in the 20 GHz, 30 GHz, and 40 GHz range are often fabricated in suspended substrate.

With the transmission media presented and briefly discussed, let us proceed to describe microstrip and stripline packaging. (Recall that we are not going to cover suspended substrate specifically; we will refer to it, however, throughout other discussions.) Each type of packaging will be presented with pointers, suggestions, and recommendations that will aid in obtaining working microwave circuits. It should be mentioned here that we will not go into detailed designs of packages but only indicate areas that should be watched when packaging microstrip and stripline and give suggestions that will aid in the proper design of a package.

Before describing specific types of packages it should be noted that every microwave package should be treated as an individual unit: That is, treat every package as a unique design that must do a particular job. Many requirements put on microwave packages result in many nonstandard arrangements. Figure 7.3 shows some of the nonstandard assemblies that are used in microwave systems. You can see that no two of them are alike; each has specific requirements to perform and must fit into a certain specified space, which dictates the shape.

Figure 7.3a Microwave packages.   Courtesy of Omni-Spectra, Inc.

Courtesy of Hyletronics Corp.

Figure 7.3b Microwave packages.

The second requirement for a microwave package, other than fitting in the right space, is for it to be very *low weight*. This is usually accomplished by building the cases out of aluminum with the walls, floor, and cover as thin as practical. This aluminum is usually either silver or gold plated to increase conduction while maintaining the low weight (a boron/nickel plating can also be used that is lower in cost and has excellent solderability).

In an effort to keep the weight of microwave packages to a minimum, and also to provide required isolation between circuits, a process called *channelizing* or *vaning* is used. Figure 7.4 shows just such a channelizing arrangement used to provide isolation between switch elements. You can see how such an arrangement would take a certain amount of weight from the package by having the circuit area milled out of the case and at the same time provide a "tunnel" (or channel) for good isolation. Care must be taken when designing such a case, however, since improper dimensions of the channels can cause unwanted modes to be set up, which will cause problems with circuit operation. Be aware of the frequencies you are using and whether the amount of isolation required is such a value to justify the time and expense of channelizing the case.

With some of the basic of cases for microwave application presented, we can now get more into specific uses: microstrip and stripline.

Figure 7.4 Channelized packages.   Courtesy of M/A-COM

7.2 MICROSTRIP PACKAGES

This type of packaging is probably one of the most difficult because it involves two different dielectric medias and all of the rules governing each of them. The two dielectric media are *air on top* (ϵ_o = 1.0) and the *circuit dielectric* ($\epsilon = \epsilon_r$) itself. Figure 7.5 shows these relationships. You can see that packaging microstrip is like putting an object in an oversized box. Part of the box is filled with the object itself, and the rest is filled with air.

The first condition that must be considered when packaging microstrip circuits is the attachment of the substrate to the case. This involves much more than simply fitting a substrate to an aluminum box. The method selected will depend, to a large degree, on the material used for the substrate because the coefficient of thermal expansion becomes an important parameter when firmly attaching a substrate to a case. This parameter, of course, is the amount of movement there is with a material during a change in temperature and is expressed as ppm/°C (parts per million per degree centigrade).

You either must match coefficients of expansion of the case and substrate or provide a *buffer* material between the two if the coefficients are different.

If, for example, we are using an aluminum case and decide to use a high-dielectric ceramic-loaded PTFE material (ϵ_r = 10.2), we would have a choice of either mounting the material to the aluminum case by means of soldering or conductive epoxy (provided the aluminum has been properly plated) or purchasing the material already bonded to an aluminum plate. The methods of attachment are this direct, since the coefficients of thermal expansion for high-K material (ϵ_r = 10.2) and aluminum are very similar. (Aluminum has a coefficient of expansion of 24 ppm and the high-K material is 20 to 25 ppm.) By attaching the substrate directly to the plated aluminum case you should not encounter any problems with breaking connection either on the substrate or the connector tabs to the outside world.

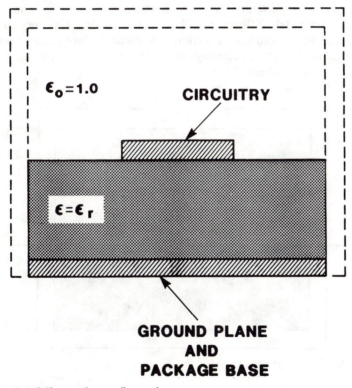

Figure 7.5 Microstrip configuration.

If, on the other hand, you choose to use an alumina substrate in an aluminum case, the task will not be as direct. The problem arises, first of all, in the coefficients of expansion. As mentioned above, the coefficient of expansion of aluminum is 24 ppm/°C. When you compare this with alumina (130 ppm/°C), you can appreciate the problem more fully. The expansion of alumina is over five times that of the aluminum case, which suggests that there is a possibility of a solder or epoxy joint's cracking if direct contact between the aluminum case and the alumina substrate is made and run over any temperature range. This is the area where a buffer metal is needed, as mentioned previously in this section, and it is usually kovar. This metal was covered in Chapter 3, and you will recall that it is a combination of iron, nickel, and cobalt (all ferrous metals). Its thermal properties (that is, its thermal coefficient of expansion, for example) are similar to those of alumina substrates. This metal will form a buffer between the alumina substrate and the aluminum case to keep solder or epoxy connections from being stressed over temperature extremes, and is thus the best method for connecting materials with widely different coefficients of expansion. (In every application you should check to be sure there are no conflicting intermetallic connections being made that will cause questionable or unreliable connection between a substrate and case.)

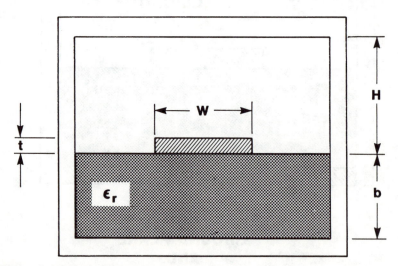

Figure 7.6 Microstrip package height.

A second condition that must be considered when packaging microstrip circuits is the can height above the circuit, designated as *H* in Figure 7.6. This is the distance that the top of the case is placed from the circuit: a critical dimension that will determine whether a final circuit will operate properly. When most microstrip circuits are breadboarded this is usually not a primary consideration, since the circuit is placed on a plate for initial testing and there is no case or top cover as such. However, when the circuit is finally packaged, the height of the cover from the circuit is a critical dimension.

To show the importance of the H dimension, the following example is presented. Suppose we have a circuit with the following parameters:

b = 0.025 inches
t = 0.0007 inches (½-oz copper)
ϵ = 10.2
w = 0.060 inches (approximately a 34- to 35-ohm line without the cover being considered)

To show how the impedance is effected and thus the circuit operation, we will concentrate on ϵ_{eff}, the effective dielectric constant. Variations in this parameter will directly effect the impedance of a microstrip line. The effective dielectric constant can be defined as:

$$\epsilon_{eff} = \frac{\epsilon_r + 1}{2} + q \; \frac{\epsilon_r - 1}{2}$$

where:

ϵ_r = the relative dielectric constant of a material as given on a manufacturer's data sheet, and

q = the filling factor that compensates for the two dielectric constants of microstrip, air and the solid dielectric of the circuit.

The quantity q is further defined as:

$$q = (q_\infty - q_t)qc$$

where:

q_∞ = the filling factor for an infinite cover height,
q_t = correction for a finite conductor thickness, and
q_c = correction for a noninfinite shielding.

You may have also seen the effective dielectric constant expressed in a different way:

$$\epsilon_{eff} = 1 + q\,(\epsilon_r - 1)$$

This is a valid expression and is used in microstrip design to calculate circuit parameters. The filling factor, q, in this case does not take into account the effect of the cover (basically q_∞) and usually has a value that is approximately twice that of the q used in our first effective dielectric constant formula. Both equations will result in the similar effective dielectric constant values. We will show this later in this section. Any variations can be attributed to equations approximations and to any rounding off that may have occurred during calculations. They should be fairly close.

With the effective dielectric constant expressed and its importance emphasized, we can proceed with the calculations of q (filling factor) that are to be used. As previously noted, there are three parts to this q: q_∞, q_c, and q_t. Working backwards with these terms they can be expressed as follows:

1.) $\quad q_T = \dfrac{2 \ln 2}{\pi} \quad \dfrac{t/b}{(w/b)^{1/2}}$

2.) $\quad q_c = \text{Tanh}\left(1.043 + 0.121\,(H/b) - \dfrac{1.164}{(H/b)}\right)$

3.) $\quad q_\infty = \left(1 + \dfrac{10b}{w}\right)^{j}$

where: $j = a\,(w/b)\,b\,(\epsilon_r)$

and

$$a\,(w/b) = 1 + \frac{1}{49}\,\ln(w/b)^2\,[(w/b)^2 + (1/52)^2]$$

$$[(w/b)^4 + 0.432] + \frac{1}{18.7}\,\ln\left[1 + \left(\frac{w}{18.1b}\right)^3\right]$$

$$b\,(\epsilon_r) = -0.564\left(\frac{\epsilon_r - 0.9}{\epsilon_r + 3}\right)^{.053}$$

(In each of the expressions above W = the strip width; t = the copper thickness; b = the dielectric thickness; H = the height of the cover above the circuit; and ϵ_r = the relative dielectric constant. Each is shown in Figure 7.6.)

You can see that the above expressions involve several calculations, but it is not the intent of this text to discuss involved mathematics. The equations are shown to illustrate what parameters (w, t, H, and b) are involved in arriving a case height: That is, q_t for the ratios of t/b and w/b; q_c for the ratio of H/b; and q_∞ for the parameters W and b with a reference made to dielectric constant (ϵ_r).

As mentioned above, the filling factor component for case height, H, is q_c. This is the logical starting point in our example to show the effects of the cover. (For this example we will vary H from 0.21 to 0.36 inches.) Figure 7.7 shows q_c for the various values of H we have assigned. If you refer back to the expression for total q, you can see that the closer you get q_c to 1.0, the less effect the cover will have on the circuit. This also can be seen in Figure 7.7.

To illustrate the full impact of the case height over our chosen values, the parameters were put into a computer program and allowed to vary. Figures 7.8 through 7.11 show the variations in W, and thus variations in impedance, caused by varying case height and filling factor, q. The indirect effect, of course, is a variation in effective dielectric constant during this process. You will note that as the cover height, H, goes beyond the 0.26 in height the curve does not change significantly. This value coincides closely with our rule of thumb that says we should place the cover 10 times the ground-plane spacing (b) away from the circuit. (10b = 0.250 inches in our case.) The chart below summarizes the results:

Height (H)	Filling Factor (q)	ϵ_{eff}	(ohms)
0.21	0.378	7.33	35.2
0.26	0.385	7.37	34.2
0.31	0.390	7.39	33.4
0.36	0.390	7.39	33.4

We presented two equations for finding effective dielectric constant and said that the value for both would be similar. We have enough information to illustrate this. The first value is:

$$\epsilon_{eff} = \frac{\epsilon_r + 1}{2} + q \; \frac{\epsilon_r - 1}{2}$$

$$\epsilon_{eff} = \frac{10.2 + 1}{2} + .390 \; \frac{10.2 - 1}{2}$$

$$\epsilon_{eff} = 7.39$$

For the second value the dielectric constant is:

$$\epsilon_{eff} = 1 + q \, (\epsilon_r - 1)$$

$$\epsilon_{eff} = 1 + .685 \, (10.2-1)$$

$$\epsilon_{eff} = 7.30$$

(The value of q used above is obtained by using Bryant-Weiss equations and obtaining a computer printout of filling factor versus impedance.)

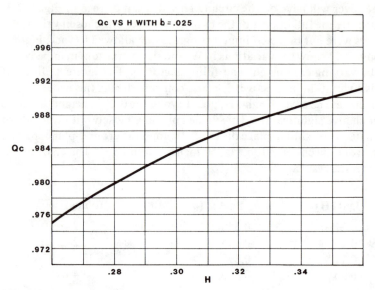

Figure 7.7 Q versus H.

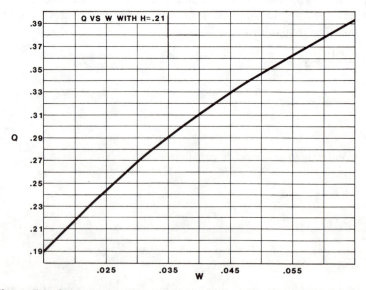

Figure 7.8 Q versus W (H = 0.21).

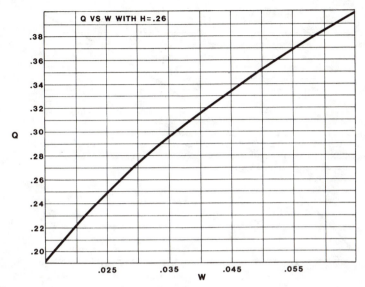

Figure 7.9 Q versus W (H = 0.26).

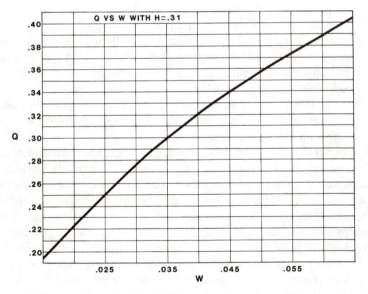

Figure 7.10 Q versus W (H = 0.31).

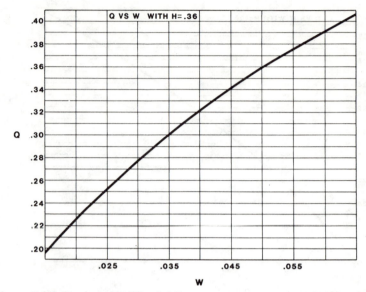

Figure 7.11 Q versus W (H = 0.36).

The value of effective dielectric constant is very close for both methods: There is about a 1 percent difference in the two methods, which has many explanations. The point of the comparison, however, is that you can use either method and still be compensated for a cover height.

A third condition that can arise in both microstrip and suspended substrate circuits is the generation of unwanted waveguide modes. As you will notice from previous figures showing microstrip and suspended substrate, they very closely resemble an end view of a piece of waveguide. For microstrip the *a* waveguide dimension would be the horizontal distance, and the *b* dimension would be the height of the cover. For suspended substrate the *a* and *b* dimensions would be the actual case dimensions with emphasis placed on the t/b (board thickness compared to overall case height) dimension, which will determine if a mode will be reinforced and thus cause problems. These relationships are shown in Figure 7.12. The analysis of modes and dimensions is beyond the scope of this text. If you are designing a microstrip or suspended substrate circuit, you should familiarize yourself with the calculations necessary for characterizing any unwanted modes to avoid trouble with the circuits when they are fabricated.

Figure 7.13 shows some typical microstrip packaging techniques that involve a variety of components integrated into one case. Figure 7.13a is a milled aluminum case that involves the use of a variety of ceramic substrates. In many cases a high-K material could also be used to integrate many of the separate substrates shown in the figure. Figure 7.13b shows a microstrip assembly in a welded aluminum housing. In many applications laser welding can be used for fabrication of such a case. TIG (tungsten inert gas) welding can also be used. This process allows you to weld the entire package shut without significantly raising the overall package temperature, provided the unit is properly heat sunk.

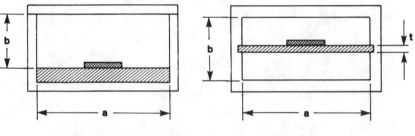

a) MICROSTRIP

b) SUSPENDED SUBSTRATE

Figure 7.12 Waveguide dimensions.

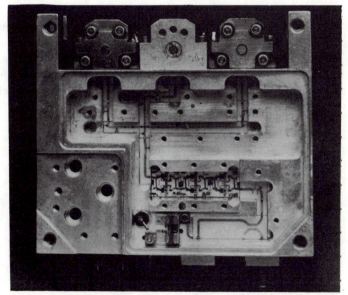

Figure 7.13a Microstrip package.   Courtesy of NARDA Microwave

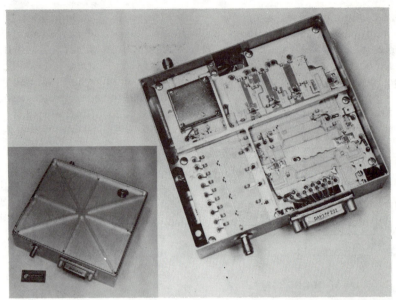

Figure 7.13b Microstrip package.   Courtesy of M/A-COM

7.3 STRIPLINE PACKAGING

Stripline packages are very different from the previously covered micro-strip cases. All that usually can be seen is a small flat block that has connectors on it and is painted a certain color depending on requirements or vendor. Such small flat blocks as shown in Figure 7.14 are common packages for stripline circuits. The objective of this section is to show how these packages are fit with the microwave laminate to form a working system.

Courtesy of Anaren Microwave

Figure 7.14a Stripline package.

Courtesy of Anaren Microwave

Figure 7.14b Stripline package.

One difference between the stripline and the previously covered microstrip packages is that there is only one dielectric to work with. Just as packaging microstrip is like putting an object in an oversized box, packaging stripline is like putting an object into a custom-fit box: There is no air or excess room involved or allowed.

There are a variety of methods used for packaging stripline circuitry. We will cover four basic types of packages and make reference to any variations that may be used: the *sandwich package* used for breadboarding, *channelized chassis*, *box and cover*, and the popular *caseless package*.

The first type package is shown in Figure 7.15 in a side view. A close look at the construction will show why it is excellent for stripline breadboarding. The package consists of an aluminum plate on top, the two dielectric pieces (or three if a thin dielectric is used for the circuit as in the case of a quadrature hybrid), and the bottom aluminum plate. This construction fulfills all of the criteria necessary for stripline operation.

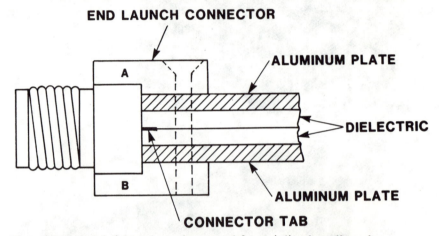

Figure 7.15 Sandwich-style package used for stripline breadboards.

The connector is different from the typical package connectors used for the type of enclosure usually associated with stripline and microwave circuits. The usual connector may have either a two- or four-hole flange that is attached to the metallic portion of the case with the center conductor attached to the circuit. In another arrangement the only part of the connector is the threaded portion that is inserted into the case, which is tapped to accept the connector and have the center conductor, once again, attached to the circuit. The connector in Figure 7.15 does neither of these things; instead it forms a clamp to hold the stripline package together and apply very tight pressure on the center conductor to hold it on the circuit. This type of connector is called an *end launch* connector. Spacings available between the top plate of the connector (A in Figure 7.15) and the bottom plate (B), are 0.062 in., 0.125 in., and 0.250 in. Those most widely used are the 0.125 in. and 0.250 in. These allow you to use two pieces of 0.030 in. PTFE/glass laminates with two pieces of 0.030 in. aluminum in one case, and two pieces of 0.062 in. PTFE/glass laminate with two pieces of 0.062 in. aluminum in the other. This connector will allow you to assemble a package for breadboard and easily disassemble it to make changes.

A variation of the stripline package covered above is called *flat plate* construction. In this type of construction a flange-mounted connector is used since the plates wrap around the circuit and allow this type of connector to be used. It is a bit more difficult to assemble and disassemble but still has an advantage of simplicity.

A second type of stripline package is the *channelized chassis*. Figure 7.16 shows a case that has been milled with channels that accept the stripline laminates. The circuits are flush with the case and are not set on top of the case itself. These cases can either be machined or stamped to the needed configuration. The paths into which the circuit will be placed are milled to preserve the wall proximity on both sides of the circuit. This type of construction is very complex and can be expensive. It therefore, generally is used only where the circuit performance necessitates and justifies the high cost.

Figure 7.16 Channelized chassis.   Courtesy of M/A-COM

A third method, called *box and cover*, is a popular type of stripline package. The circuits shown back in Figure 7.14 are all fabricated using this type of construction. The method is similar to the channelized chassis in that the board is cut. In the box and cover chassis, however, the board is not cut to conform to the circuit pattern but to the outside edge (border) of the board. In this way the circuit is dropped into the case as before, but this method does not involve the complexity of cutting, as in the case of the channelized chassis. This is shown in Figure 7.17. You can see that the "box" and "cover" are each machined so that the circuit board (laminate) can be recessed into the metal itself, resulting in a firm, tight fit of the stripline package, which probably explains why it is so widely used througout the microwave industry where stripline is fabricated.

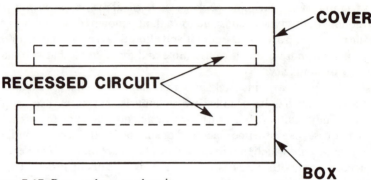

Figure 7.17 Box-and-cover chassis.

Figure 7.18 shows how connectors can be put in the box-and-cover type of case. Generally the barrel of the connector is extended from the conventional type connector so that it can be inserted into the case without having threads inside the metal portion of the case. A slot (shown in the figure) is cut in the channel to accommodate the flange portion of the connector (the flange is equivalent to the two-hole flange size). This machining arrangement gives the connector a solid place to set itself and thus eliminates any twist (or torque) on the connector that could break the connector tab away from the circuit.

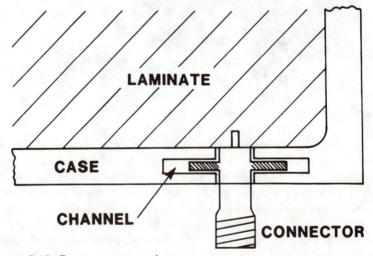

Figure 7.18 Connector mounting.

The final type of stripline package to be covered is called the *no chassis* or caseless package. This type package is useful when drop-in components are to be used. Figure 7.19 shows two close-up views of such a case. Figure 7.19a shows the package with tabs coming out the edges for connection to other circuits. These are shown in the foreground of Figure 7.19b. The rivets are used for two purposes. The small ones simply ensure that the package remains together. The large ones accomplish the same task but also are used to mount the component to a larger chassis and, at the same time, ensure a good ground connection.

Figure 7.19 Caseless package.

The package is assembled by placing the two ground-plane laminates together with the thin dielectric material between them. Aluminum plates are placed on the top and bottom of the ground-plane laminates, and heat and pressure are applied (thermocompression process) until the package is fused together. Sometimes an adhesive or a bonding film (such as 3M No. 6700) is placed between layers prior to the thermocompression phase.

The obvious advantages of such a package are that no complex machining is necessary and that no connectors are involved to cause losses or excessive VSWR's. One large disadvantage is that the unit is not repairable. If a problem arises, the unit must be scrapped and replaced with a new one. You will completely tear the laminates apart before you even begin to separate them, if you are so inclined to try. Do not use this package for active devices unless you intend to make it a throw-away module. Also, be sure that your active devices will take the heat involved in the assembly of such a case; most active devices will not take the lamination temperatures. Thus, the caseless type chassis seems to be limited to passive devices.

There are a variety of case styles that can be used for stripline circuits, depending on your particular application. Take great care in choosing the style you use, since it can either make or break your carefully built microwave design. (Additional circuit packages are shown in Figure 7.20.)

Figure 7.20a Stripline package.   Courtesy of M/A-COM

Figure 7.20b Stripline package.

7.4 PACKAGE SEALING

The packages covered in this chapter for stripline, microstrip, and suspended substrate generally have to protect the internal circuit from elements such as moisture, humidity, salt spray, dust, and outside electrical interference (EMI). In order to protect the circuit adequately certain methods of *sealing* can be used. Some of these methods are presented below:

- *Conductive epoxy*: This is an excellent sealer that can be put around the outside edge of the case and cured to make the box essentially one piece. This method of sealing generally makes the assembly non-repairable.
- *Solder*: This is a very effective method of sealing. Low-temperature solder should be used so that repair can be made on the package if necessary.
- *Gasket material*: This can be placed in the case as it is put together to form an excellent seal. Care should be taken to ensure the material is not too thick, which could upset the ground-plane spacing of the stripline circuit.
- *Welding*: This is becoming more popular as a means of sealing microwave packages. It is very effective and also allows for a certain number of repairs.

- *Metallized tape*: This is an inexpensive method of sealing a microwave package. Commercial applications will use such a method. It is available in both conductive and nonconductive adhesive and can also utilize thermosetting adhesives that can be baked, resulting in even better sealing.

7.5 CHAPTER SUMMARY

This chapter has covered a very important step in the fabrication of microwaves: packages. Microwave packages are much more than just a box to put a circuit in: They are an integral part of the overall circuit design. When your ultimate circuit design is completed, take at least as much time working on the package as you did on the circuit. It will pay off handsomely in the long run.

CHAPTER 8

CONNECTORS AND TRANSITIONS

8.1 INTRODUCTION

Our discussions thus far have concentrated on the microwave circuit itself and putting it into the proper case. We have investigated materials to use; generated artwork for the circuit; etched the final circuit; assembled the circuit using the appropriate bonding method; and placed the finished circuit into a package. One step remains before obtaining an actual working microwave circuit: In order to be able to get signals into and out of the circuit, there must be connectors on the case and transitions from the finished circuit to these connectors.

8.2 MICROWAVE CONNECTORS

A variety of connectors are available today referred to by acronyms such as SMA, SMB, SMC, SC, C, BNC, TNC, N, LT, HN, QDS, QC, TPS, APC 3.5, APC-7, and so on. Some of these connectors are shown in Figure 8.1. These letters and numbers describe connectors that have many different and widely varied applications. The one that is best for your circuit depends on what your circuit has to do and what frequency range it has to operate over. The microwave connectors covered in this chapter are the SMA, TNC, type N, and APC-7; these are the primary connectors used in microwaves. You will see how the dimensions of these connectors make them suitable for the higher-frequency region of microwaves.

Figure 8.1 Microwave connectors.   Courtesy of Omni-Spectra, Inc.

The microwave connector should be considered a separate microwave component and as such should be given the same extensive consideration required when choosing a directional coupler, for example, for a specific test setup. Never underestimate the importance of connectors to the operation of your circuit. Many times a circuit is assumed to have a problem when a bad connector on the case is keeping the circuit from operating properly. In this case, simple connector replacement will cause all systems to operate to their expected level of performance and allow you to go on to new and exciting challenges.

When we say the microwave connector is a separate component, it means that the microwave connector has certain specifications by itself, just as any other component. For the microwave connector these specifications are such areas as characteristic impedance, upper frequency limit, CW power capability, size and weight, VSWR, environmental conditions (vibration, moisture, and temperature), field replaceability, contact characteristics, and cost. If the connector is to be used for lab applications, you should also consider its life: How many connect/disconnect cycles can it handle? A connector is much more than a piece of metal with threads on it; it is a component unto itself.

Specifications that govern microwave connectors are presented in MIL-C-39012. As of this writing the specification is in revision C and consists of 116 separate specification sheets. A supplement sheet contains a description of the connector (cabled plug, male, pin, or straight, for example), a picture of the body style, where the connector can be found within the overall specification, and the applicable cables to be used with that particular cable.

The specification describes two classes of connectors and five categories:

- *Class 1*: A class 1 connector is intended to provide superior RF performance at specified frequencies, and all its RF characteristics are completely defined. (These are connectors to be used in critical applications.)
- *Class 2*: A class 2 connector is intended to provide mechanical connection within an RF circuit providing specified RF performance. (These are used for less critical applications.)
- *Category A*: Field-serviceable connectors.
- *Category B*: Non–field-replaceable connectors.
- *Category C*: Field-replaceable solder center contact.
- *Category D*: Field-replaceable crimp center contact.
- *Category E*: Field replaceable.

(For specific requirements you should refer to MIL-C-39012 Rev C dated 11 August 1982. This should define all necessary parameters for any microwave connectors available.)

The parameters mentioned above are all important in choosing the right connector for your particular application. Some may apply directly, others may not apply at all. Each should be considered, since they all may have some bearing on circuit operation. All these parameters are determined by the physical size and construction of the connector itself. There are also times when the circuits you are connecting to will effect which connector you will use. Additional considerations for deciding which connector is appropriate include:

- *Cable size*: The best electrical performance can be achieved if the connector is approximately the size of the cable to be used. Every time there is a change in dimension (diameters, for example), there is a possibility of discontinuities; these will have to be compensated for in order to maintain a good match in the transition and ensure low loss.
- *Frequency range*: The frequency range in which a connector can be used will decrease as the size of the connector increases. This you can see with a type N connector (0.630 in diameter) is rated to approximately 12 GHz while an SMA connector (0.260 in diameter) will operate up to 18 GHz. A connector between these is the TNC (0.440 in diameter) which is rated to 15 GHz.

- *Connector coupling method*: This is how mating connectors of the same type are connected. Most connectors are threaded, although some have a bayonet-type coupling used for quick disconnect. Until recently, when many system requirements dictated small, compact packages, the quick-connect/disconnect type connector was not acceptable. However, recent developments in connector design have made this type coupling more widely used for a variety of applications.
- *Contact captivation*: Captivation occurs when the center conductor of the connector is secured within the connector so that there is no axial movement due to temperature or cable flexing. Generally, the captivated connector is used where good mechanical requirements and general electrical performance is needed. Where the electrical performance takes precedence over the mechanical aspects, the noncaptivated connector is preferred.

At this point you should realize the importance of microwave connectors and why they should be treated as separate components. Careful consideration must be given to choose, not only *a* connector but the *right* connector. To aid in this decision we will now present brief descriptions of the SMA, TNC, type N, and APC-7 connectors.

8.2.1 SMA Connectors

The SMA connector (specified in MIL-C-39012/55 through MIL-C-30912/62, MIL-C-39012/79 through MIL-C-39012/83, and MIL-C-39012/92 through MIL-C-39012/94) is probably the most widely used connectors in microwaves. Figure 8.2 shows a picture of a typical male/female connector set in SMA configuration. Figure 8.3 is a hermetic version of the SMA that is completely threaded and will turn into a mating threaded area of a case.

The SMA comes in a variety of configurations. Figures 8.2 and 8.3 show two of these. Other types of SMA connectors are listed below. (All designations referred to as *jacks* are male connectors and as *plugs* are female.)

- Straight cable plug
- Straight cable jack
- Bulkhead feedthrough cable jack
- Flange-mount cable jack
- Two-hole flange-mount cable jack
- Right-angle cable plug
- Swept right-angle cable plug
- Flange-mount jack receptacle
- Flange-mount plug receptacle
- Bulkhead-feedthrough jack receptacle
- Printed circuit board straight and right-angle jack and plug
- End-launch jack and plug stripline

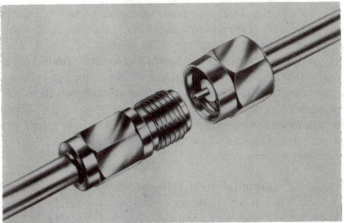

Courtesy of Omni-Spectra, Inc.

Figure 8.2 SMA connectors.

Figure 8.3a SMA hermetic connector.

Courtesy of Omni-Spectra, Inc.

Figure 8.3b SMA hermetic connector.

There is a wide variety to choose from when using SMA connectors. This variety exists because of the wide usage of SMA's and the varied requirements imposed on connector manufacturers by designers.

Figure 8.4 shows dimensions for the SMA connector. Both the male (jack) and female (plug) are shown. Notice the small dimensions (0.260 in. diameter for the male and 0.182 in. diameter for the female), which make this an excellent high-frequency device.

If you were to look in a connector catalog for SMA connectors, you would see typical electrical specifications as shown below:

Requirement	Specification
VSWR	From DC to 18 GHz the VSWR shall not exceed 1.02 + 0.005 (f), where f = frequency in GHz
RF leakage	RF leakage shall not exceed –(100-f) dB, where f = frequency in GHz
Insertion loss	The insertion loss in 0.03 $\sqrt{f}$ dB max, where f = frequency in GHz

These specifications are saying that:

- The VSWR of the connector at 10 GHz, for example, is 1.02 + 0.005 (10) or 1.07 max.
- The RF leakage of the connector at 10 GHz will be –(100 – 10) dB or –90 dB max.
- The insertion loss of the connector at 10 GHz will be 0.03 $\sqrt{10}$ dB or 0.0948 dB max.

Mechanical specifications for the connector would be:

- Force to engage and disengage (2 in.-lb max)
- Coupling-nut retention force (15 in.-lb min)
- Cable retention force (60 lb min)
- Mating characteristics
- Connector durability

Environmental specifications may be as follows:

- Vibration
- Shock
- Thermal shock
- Recommended mating torque (7 to 10 in.-lb)
- Corrosion (salt spray)
- Moisture resistance

The SMA connector is the work horse connector for the microwave industry. Applications for this tiny connector seem to be unlimited.

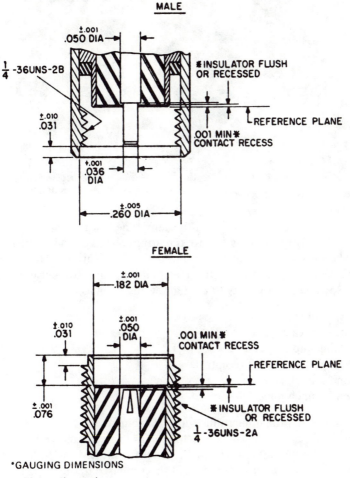

Figure 8.4 SMA dimensions.

8.2.2 TNC Connectors

The TNC connector is the next larger size up from the SMA previously discussed. The TNC is covered in MIL-C-39012/26 through MIL-C-39012/34 and MIL-C-39012/112 through MIL-C-39012/116.

Figure 8.5 shows the TNC connector. It actually is a BNC connector with threads. The BNC is the connector used in many lower-frequency applica-

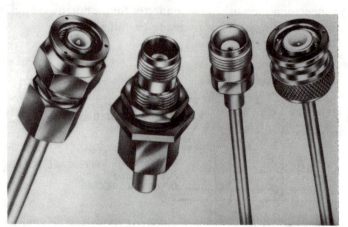

Figure 8.5 TNC connectors.   Courtesy of Omni-Spectra, Inc.

tions. You have undoubtedly seen it many times on the ends of a piece of
RG-58 cable. The BNC twists on to its mating connector and, as such, would
not make an adequate ground for microwave applications. Thus, the TNC
uses the features of a BNC with threads on it for an excellent ground. The
TNC also comes in a variety of configurations, such as:

- Straight cable plug
- Straight cable jack
- Flange-mount cable jack
- Flange-mount jack receptacle
- End-launch jack
- End-launch plug

Figure 8.6 shows the dimensions for TNC connectors. It is larger than the
previously discussed SMA connector: a 0.440 in. diameter for the male
connector as opposed to a .260 in. diameter for the SMA. The TNC female
connector also has a diameter of 0.381 in. compared to the 0.182 in. dimension for SMA's. Typical electrical specifications for a TNC connector are
shown below.

Requirement	**Specification**
VSWR	From DC to 15 GHz the VSWR shall not exceed 1.07 + 0.007 (f), where f = frequency in GHz
RF leakage	Shall not exceed −(90-f) dB, where f = frequency in GHz
Insertion loss	Shall be 0.05 $\sqrt{f}$ dB max, where f = frequency in GHz

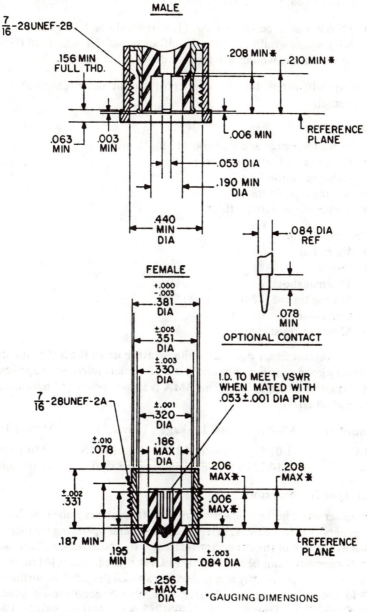

MALE

$\frac{7}{16}$-28UNEF-2B

.156 MIN
FULL THD.

.208 MIN ✳   .210 MIN ✳

.063
MIN

.003
MIN

.006 MIN

REFERENCE
PLANE

.053 DIA

.190 MIN
DIA

.440
MIN
DIA

.084 DIA
REF

.078
MIN

OPTIONAL CONTACT

FEMALE

+.000
-.003
.381
DIA

±.005
.351
DIA

±.003
.330
DIA

I.D. TO MEET VSWR
WHEN MATED WITH
.053 ±.001 DIA PIN

$\frac{7}{16}$-28UNEF-2A

±.001
.320
DIA

±.010
.078

.186
MAX
DIA

.206
MAX ✳

.208
MAX ✳

±.002
.331

.006
MAX ✳

.187 MIN

REFERENCE
PLANE

.195
MIN

±.003
.084 DIA

.256
MAX
DIA

✳GAUGING DIMENSIONS

Figure 8.6 TNC dimensions.

Values for the parameters above for 10 GHz would be as follows:

- VSWR would be 1.07 + 0.007 (10), or would be 1.14:1 max.
- RF leakage would be –(90 – 10) dB or be –80 dB max at 10 GHz.
- Insertion loss would be 0.05 $\sqrt{10}$ dB or 0.158 dB max.

Other specifications for the connector would be mechanical and environmental:

Mechanical
- Force to engage and disengage (2 in.-lb max)
- Coupling-nut retention force
- Cable retention force
- Mating characteristics
- Connector durability (life)

Environmental
- Vibration
- Shock
- Thermal shock
- Mating torque (12 to 15 in.-lb)
- Corrosion (salt spray)
- Moisture resistance

The TNC connector does not quite measure up to the SMA but still has some very good specifications and is used for many microwave applications. A direct comparison of TNC and SMA is shown below (all parameters are shown at 10 GHz):

Connector	VSWR	RF Leakage	Ins. Loss	Mating Torque
SMA	1.07:1	–90 dB	0.0948 dB	7 to 10 in.-lb
TNC	1.14:1	–80 dB	0.158 dB	12 to 15 in.-lb

8.2.3 Type N Connectors

Another work horse at lower microwave frequencies is the type N connector, generally used below 12 GHz. It is larger, and operating frequency goes down as the size of the connector increases. By special design, however, the type N connector can be used up to 18 GHz if used with 0.141 in. semirigid cable. These, as previously mentioned, are special design. The normal type N should not be used above 12 GHz. The type N connector is covered in MIL-C-39012/1 through MIL-C-39012/5 and MIL-C-39012/117 (EC) through MIL-C-39012/121 (EC) with the latter sections used with semirigid cables.

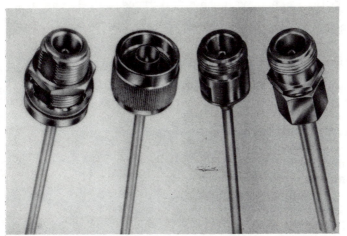

Figure 8.7 Type N connectors.   Courtesy of Omni-Spectra, Inc.

Figure 8.7 shows a picture of some type N connectors. Notice the larger size of these connectors. This is confirmed in Figure 8.8, which is a mechanical drawing of the device. Notice that the diameter of the male connector, for example, is 0.630 in. as compared to 0.44 in. for TNC and 0.260 in. for SMA. Also, the diameter of the female connector is 0.627 in. as compared to 0.381 in. for TNC and 0.182 in. for SMA.

This connector is available in a variety of configurations, just as the SMA and TNC are. Some typical configurations are:

- Straight cable plug
- Straight cable jack
- Flange-mount cable jack
- Straight-panel cable jack
- Flange-mount plug/jack receptacle
- End-launch plug/jack

Specifications for the type N connector are shown below for electrical parameters:

Requirement	Specification
VSWR	From DC to 10 GHz the VSWR shall not exceed 1.06 + 0.007 (f), where f = frequency in GHz
RF leakage	Shall not exceed –(95-f) dB, where f = frequency in Ghz
Insertion loss	Shall be 0.05 $\sqrt{f}$ dB, where f = frequency in GHz

The three electrical parameters above calculated at 10 GHz are as follows:

- VSWR for the connector will be 1.06 + 0.007 (10) or 1.13:1 max.
- RF leakage will be –(95-10) dB or –85 dB max.
- Insertion loss will be 0.05 $\sqrt{10}$ dB or 0.158 dB max.

Other specifications for the type N will be mechanical and environmental:

Mechanical
- Force to engage and disengage (3 in.-lb)
- Coupling-nut retention force
- Cable-retention force
- Mating characteristics
- Connector durability (life)

Environmental
- Vibration
- Shock
- Thermal shock
- Recommended mating torque (12 to 15 in.-lb)
- Corrosion (salt spray)
- Moisture resistance

We indicated in the beginning of this section that the type N connector was a work horse for lower microwave frequencies. This can be seen by comparing the electrical parameters of the type N connector to TNC and SMA (the reference is, once again, 10 GHz):

Connector	VSWR	RF Leakage	Ins. Loss	Mating Torque
SMA	1.07:1	–90 dB	0.0948 dB	7 to 10 in.-lb
TNC	1.14:1	–80 dB	0.158 dB	12 to 15 in.-lb
N	1.13:1	–85 dB	0.158 dB	12 to 15 in.-lb

You can note immediately how the type N would be good for lower microwave frequency applications, since at 10 GHz it is as good or better than the TNC. To illustrate this even further the type N connector would perform as follows at 5 GHz:

- VSWR = 1.095:1
- RF leakage = –90 dB max
- Ins. loss = 0.111 dB max

The type N connector does the same excellent job at lower microwave frequencies that the SMA does at the higher frequencies.

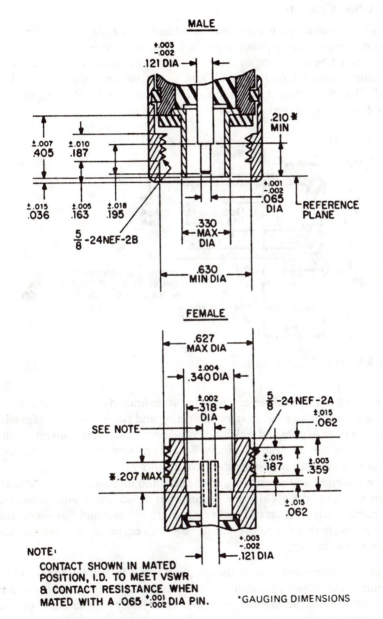

NOTE:
CONTACT SHOWN IN MATED
POSITION, I.D. TO MEET VSWR
& CONTACT RESISTANCE WHEN
MATED WITH A .065 $^{+.001}_{-.002}$ DIA PIN.

*GAUGING DIMENSIONS

Figure 8.8 Type N dimensions.

8.2.4 APC-7 Connectors

All of the connectors presented thus far have had a pin in one connector that mated with a corresponding pin in its partner connector. The APC-7 makes contact by a pressing of the two connector assemblies together. This connector is also referred to as a 7mm connector and is not covered in MIL-C-39012C; rather, it is found in IEEE Standard No. 187. Figure 8.9 shows dimensions of the connector.

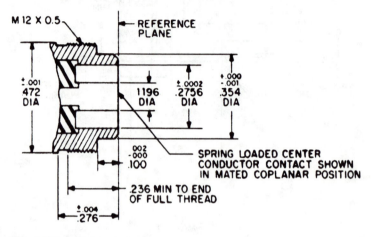

Figure 8.9 APC-7 dimensions.

The 7mm (APC-7) connector comes in rather limited configurations compared to other connectors: straight cable plug and flange-mount receptacle. The designation of plug and jack is not valid here, since there is no pin on the connector, as can be seen in Figure 8.9. When you put two connectors together you are mating two identical connectors.

There is a right and wrong way to connect and disconnect these devices. Figures 8.10(a) and 8.10(b) show the internal construction and connector parts, respectively. It is crucial that you learn to connect and disconnect the APC-7 properly in order to prevent damage and ensure long, accurate performance. To connect:

1. On one connector, retract the coupling sleeve by turning the coupling nut counterclockwise until the sleeve and nut disengage. This point is reached when the coupling nut can be spun freely with no motion of the coupling sleeve.

2. On the other connector, fully extend the coupling sleeve by turning the nut clockwise. Once again, the coupling nut can turn fully when the

sleeve is fully retracted. To engage the coupling sleeve and coupling nut when the sleeve is fully retracted, press back lightly on the nut while turning it clockwise.

3. Put the connectors together firmly, and thread the coupling nut of the connector with the retracted sleeve over the extended sleeve. Leave the other nut in its original position; closing the gap between coupling nuts tends to loosen the electrical connection.

To disconnect:

1. Loosen the coupling nut of the connector showing the wide gold band (the one that had the coupling sleeve fully retracted when connected), as shown in Figure 8.10(b).

2. Important: Part the connectors carefully to prevent striking the inner conductor contact.

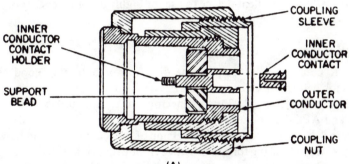

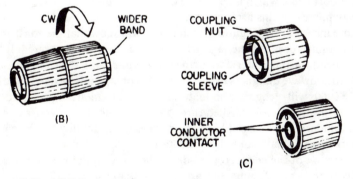

Figure 8.10 APC-7 connection.

The area where APC-7 (7mm) connectors find their widest application is in test instruments and lab test setup because they are excellent connectors to be used up to 18 GHz and are simple to correct and disconnect when the proper procedure is used. The data below show the specification limit for an APC-7 connector and measured results:

Frequency	Specification Limit	Measured Data
4 GHz	1.011	1.004
6 GHz	1.015	1.006
8 GHz	1.019	1.010
10 GHz	1.023	1.014
12 GHz	1.027	1.016
14 GHz	1.031	1.014
16 GHz	1.035	1.020
18 GHz	1.039	1.025

The APC-7 is an excellent device to use for highly accurate testing of microwave systems and components.

8.2.5 K Connector

The K connector is a step in the direction of obtaining good coaxial connector peformance above 40 GHz. The K connector is neither an SMA nor an APC-3.5; it is a 2.92mm connector that is compatable with both the SMA and APC-3.5 connectors.

Figure 8.11 shows the *spark plug* version of the K connector and how it will interface with a housing for a microwave circuit. The right side of the figure is the spark plug portion of the connector, and the left side is the housing and the glass support bead, which is actually a part of the connector. Figure 8.12 is a cut-away picture of this arrangement.

A close inspection of Figures 8.11 and 8.12 will show that the construction of the K connector is not the same as previous connectors presented: It has a support bead for the center conductor that is made of PPO (polyphenelene oxide). Microwave designers doing microstrip or stripline design in the mid- and late 1960s will recognize this material as one on which many circuits were built. This bead has holes drilled through it to reduce its dielectric constant from a standard 2.55 to 2.0. The bead provides great stability for the connector laterally and axially and also for rotational movement.

The portion of the connector inside the housing contains a 10 mil center conductor on 10 mil alumina microstrip. A bead of Corning 7070 glass along with this microstrip configuration provides good RF performance and excellent mechanical strength. Figure 8.13 shows the dimensions for the spark plug and male K connectors. Notice the compatible dimensions with the SMA covered previously in this chapter.

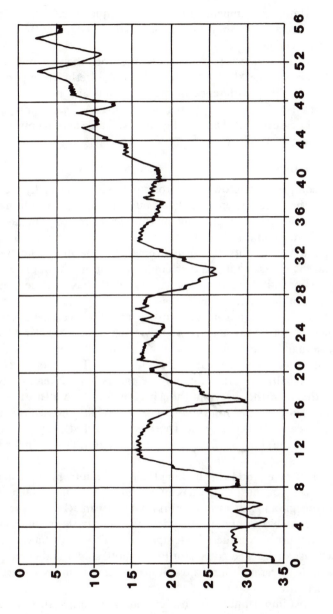

Figure 8.14 Return loss for K connectors.

exceed the maximum recommended tightening torque specification: It is a maximum value for a reason; beyond that point you will damage the connector mechanically and electrically.

Finally, keep your connectors clean, especially when they are put on and taken off frequently in the lab. A little alcohol or tricholrethylene on a cotton swab makes a world of difference in the short run and also makes the connector work better for a longer period of time. A final comment on connectors: Take care of them because they are a valuable part of your overall microwave system.

8.3 TRANSITIONS

When you complete a microwave microstrip design, you usually are convinced that it is the most efficient circuit ever built and that it is the ultimate in design. This may be simply ego talking, or your design may be outstanding. Regardless of how good it is, it is useless until you can make the transition from the circuit to the outside world. Remember the man who built a beautiful boat in his basement and then couldn't get it out the door? He had one of the most beautiful creations the nautical world has ever seen, but he couldn't get it into the water.

The problem of transition from a microstrip circuit to a coaxial connector is one that engineers have approached from every conceivable angle: Some do it over a broadband; some have a model of empirically derived elements; while others simply come up with an equivalent circuit. The list of articles on this subject is virtually endless. Though these papers have different ways of approaching the topic, they have one thing in common: They all are theoretical models of what the transition *should* look like. What is needed is a list of transitions that can be practically incorporated into a design under various conditions so that the glorious circuit you have created will blossom for all to see.

When considering methods for making the transition from microstrip to coax, or vice versa, you must realize it is much more than simply attaching the connector to the line and getting out (or in) what you want. (In certain cases it does prove to be this easy, but only after many factors have been considered.) One factor that must be considered is temperature. This single factor can be the greatest cause of cracked connector joints (both solder and epoxy) and actually cracking of ceramic substrates. It is neglected in many papers that discuss the theoretical models.

The single most important item when considering temperature effects of substrates, laminates, and their package is their *coefficient of thermal expansion*: how much things move when they get hot or cold. Table 8.1 shows various materials with their coefficients of expansion. You can see from this table how certain combinations will cause problems. The solution to these problems is either to match metals and materials or to have a transition that will compensate for the difference in expansion. We will cover the compensation aspects first.

Table 8.1
Coefficients of Thermal Expansion

Material	Expansion ($\times10^{-6}/°C$)
Aluminum	24.30
Copper	16.56
Brass	18.72
Kovar	5.04
Alumina	7.30
E-10	20-25
Woven Teflon® fiberglass	9.00

One of the first methods of compensating for differences in expansions is shown in Figure 8.15. In this figure the substrate and connector are attached by means of a flexible ribbon lead. This lead forms a sort of spring that allows each material to move around at will without affecting any other material around it. This seems like the ultimate mechanical solution. It is an excellent mechanical solution but a nightmare electrically. This ribbon adds inductance to the circuit that is very difficult to tune out. The effects of this inductance are shown in Figure 8.16. Figure 8.16(a) shows a bandpass filter with the connectors soldered directly to the substrate. You can see a well-defined band and high- and low-frequency skirts. In Figure 8.16(b) the flexible ribbons were used, and the high-frequency end of the response is disrupted. This hump in the high end of the response was never completely tuned out. Thus, the flexible ribbon, while being excellent mechanically,

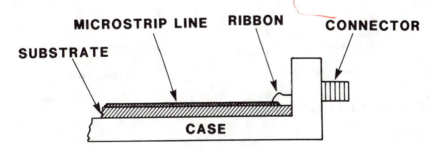

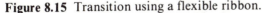

Figure 8.15 Transition using a flexible ribbon.

causes problems electrically that may not always be completely eliminated by tuning. If your circuit is not critical and you can tolerate the additional inductance, however, it may be the best method for you.

A second method allows the tab on the connector to move so that variations may be compensated for. This tab may look like the one in Figure 8.17. If you have ever put together track on a model train set, you may recognize this tab as being similar to the metal pieces used to keep the track together. This form of tab would make good electrical connection to both the connector and circuit, as shown, and would also be allowed to move freely on the connector tab.

A third method uses a flexible epoxy. One such epoxy is Epoxy Technology's H20F. This is a two-part epoxy that is designed specifically for flexible-type circuitry but can be used for transition joints where temperature may be a problem. Curing temperatures for this epoxy are similar to those of nonflexible epoxy:

- 20 minutes @ 120°C
- 60 minutes @ 100°C

H20F epoxy, from Epoxy Technology, Inc., has been put on a ceramic substrate and put through 10 cycles of −65°C to +125°C with no physical change noted. This should give a user a high degree of confidence in the material if epoxy is right for your particular application.

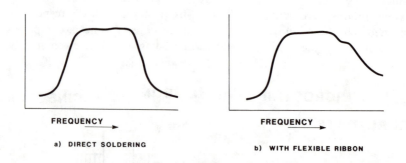

Figure 8.16 Effects of the flexible ribbon.

A fourth method for compensation for temperature in your transitions is to investigate the case design. Use a computer analysis of materials and stresses over temperature to determine where to attach substrates, or carrier plate if used, to the case to obtain the optimum results. Tests have been run on various sizes of substrates with mountings placed directly under the connector, in the center of the substrate, and on the edges of the substrate. The only conclusive results obtained thus far are that the smaller substrates (0.5 in x 0.5 in) are better over temperatures than larger ones, regardless of where they are mounted. These results, which are inconclusive as mentioned, should be taken as a starting point, and you should give careful consideration to your case design if you have any temperature range requirements.

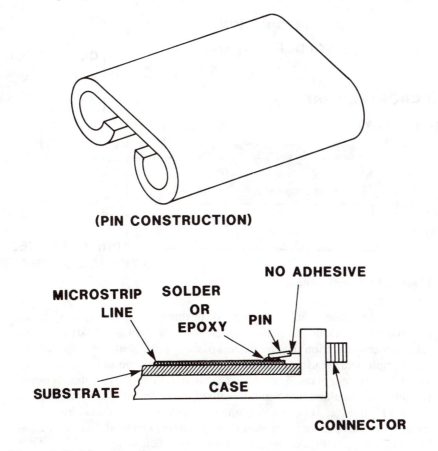

(PIN CONSTRUCTION)

Figure 8.17 Movable tab.

A new method of making the microstrip to coaxial transition is a hermetic connector that has a sliding contact on it similar to one we discussed earlier. This is along the lines of the K connector described in Section 8.2.5. Figure 8.18 shows a typical connector of this type. You can see from the figure that the heart of the connector is a hermetic seal that has the pin extending through. The pin on the right side fits into the external connector housing to make contact with the outside world. The pin on the left is attached to a pin that slides onto the seal pin and solders to the microstrip circuit. Any variation in temperature will be compensated for with the pin moving back and forth. These connectors are fairly new on the market but should solve many problems.

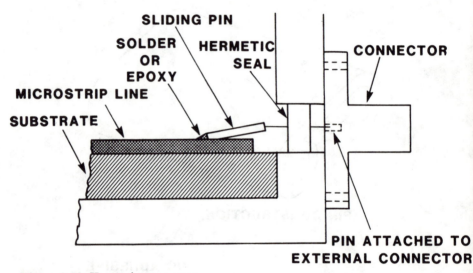

Figure 8.18 Hermetic connector.

The final method to be covered here is the one that we would like to do on all circuits: direct soldering. This can be done, as mentioned earlier, if you take into consideration the coefficients of expansion of all the materials. If, for example, you are using alumina substrates in an aluminum case, you have a drastic difference in coefficients of expansion. A kovar carrier plate between the alumina and aluminum will take care of this difference very well. Also, if you use E-10 material in an aluminum case, you will have no problem at all with expansion differences because they are almost identical. Careful matching of materials can make the job of transition much easier.

We have covered six methods of making the transition from microstrip to coax. These are not the only methods but are a representative sample. One, or several, may be right for your application. Possibly none of them will be useful to you, but you may be able to use some of the reasoning used to compile this list of six. Whatever way is best for you, be sure that when it comes time to put your glorious masterpiece in a case that you don't forget to consider the transition. Remember the man with the boat: Plan ahead, and give yourself smooth sailing.

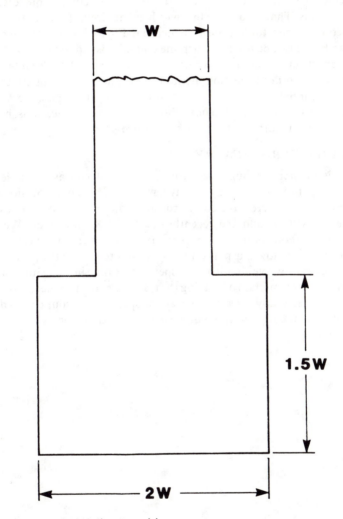

Figure 8.19 Transmission line transition.

We concentrated on coax to microstrip transitions above because this is where 99 percent of all problems occur, but consideration also should be given to stripline to coax transitions. These are also of great importance in making microwave circuits work properly. Generally, a solder connection or a press connection within the stripline will result in excellent performance of the circuit.

One final area should be covered before concluding our discussion on transitions. That area is the line width as the circuit (*stripline* or *microstrip*) meets the connector or transition medium. Figure 8.19 shows a general rule of thumb that can be used for stripline and microstrip to aid in matching your transmission line inputs/outputs to the connector, flexible ribbon, movable tab, or any other transition method used. The dimensions states that (W = width of input/output line, 2W = width of matching pad, 1.5W = depth of matching pad) may need to be tailored for specific applications, but these are good starting points in the transition process.

8.4 CHAPTER SUMMARY

This chapter, although the actual area it involved was small, tied together the state-of-the-art circuit you may have designed and the outside world. We discussed the microwave connector and concentrated on the SMA, TNC, type N, APC-7, and the recently developed K connector. We suggested various methods of transition from these connectors to the circuit.

You should now grasp how important it is to choose the proper connector for your microwave circuit and to spend some time investigating the best, and most efficient, method of making the transition from connector to circuit and vice versa. In many cases, the successful operation of your circuit depends on the degree of consideration you have given to these topics.

APPENDIXES

APPENDIX A
DIELECTRIC CONSTANTS

Material	Dielectric Constant
Air	1.00
Teflon® (PTFE)	2.10
Polyethylene	2.25
Rexolite No. 1422	2.55
Neoprene rubber	4.00
Balsawood	1.30
Snow (freshly fallen)	1.20
Snow (hard packed)	1.50
Water (distilled)	76.70

APPENDIX B
COEFFICIENTS OF EXPANSION

Element	Expansion (PPM)
Aluminum	24.3
Carbon	0.6–4.3
Chromium	6.2
Cobalt	12.3
Copper	16.56
Gold	14.2
Indium	33.0
Iron	11.7
Lead	28.7
Magnesium	25.2
Nickel	13.3
Silver	18.9
Tin	23.0
Titanium	8.5

PPM = Parts per million

APPENDIX C
CHEMICAL SYMBOLS

Element	Symbol
Aluminum	Al
Beryllium	Be
Boron	B
Carbon	C
Chromium	Cr
Cobalt	Co
Copper	Cu
Gallium	Ga
Germanium	Ge
Gold	Au
Indium	In
Iron	Fe
Lead	Pb
Magnesium	Mg
Nickel	Ni
Platinum	Pt
Silicon	Si
Silver	Ag
Tin	Sn
Titanium	Ti
Zinc	Zn

APPENDIX D
MELTING POINTS

Material	Melting Point (°C)
Aluminum	660
Cobalt	1495
Copper	1083
Gold	1063
Indium	156
Iron	1535
Lead	327
Nickel	1453
Silver	960
Tin	232

APPENDIX E
TEMPERATURE CONVERSION

Use the tables as follows:

- Find your temperature in the center column.
- If the given temperature is °C, the value of °F can be found in the right-hand column.
- If the given temperature is °F, the value of °C can be found in the left-hand column.

- *Example 1:* Given 650°C, the converted value is 1202°F.
- *Example 2*: Given 230°F, the converted value is 110°C.

Table A.E.1

C	−459.4 to 0	F
−273	−459.4	
−268	−450	
−262	−440	
−257	−430	
−251	−420	
−246	−410	
−240	−400	
−234	−390	
−229	−380	
−223	−370	
−218	−360	
−212	−350	
−207	−340	
−201	−330	
−196	−320	
−190	−310	
−184	−300	
−179	−290	
−173	−280	
−169	−273	−459.4
−168	−270	−454
−162	−260	−436
−157	−250	−418

C	0 to 100	F	C		F
−17.8	0	32	10.0	50	122.0
−17.2	1	33.8	10.6	51	123.8
−16.7	2	35.6	11.1	52	125.6
−16.1	3	37.4	11.7	53	127.4
−15.6	4	39.2	12.2	54	129.2
−15.0	5	41.0	12.8	55	131.0
−14.4	6	42.8	13.3	56	132.8
−13.9	7	44.6	13.9	57	134.6
−13.3	8	46.4	14.4	58	136.4
−12.8	9	48.2	15.0	59	138.2
−12.2	10	50.0	15.6	30	140.0
−11.7	11	51.8	16.1	61	141.8
−11.1	12	53.6	16.7	62	143.6
−10.6	13	55.4	17.2	63	145.4
−10.0	14	57.2	17.8	64	147.2
−9.4	15	59.0	18.3	65	149.0
−8.9	16	60.8	18.9	66	150.8
−8.3	17	62.6	19.4	67	152.6
−7.8	18	64.4	20.0	68	154.4
−7.2	19	66.2	20.6	69	156.2
−6.7	20	68.0	21.1	70	158.0
−6.1	21	69.8	21.7	71	159.8
−5.6	22	71.6	22.2	72	161.6

°C		°F
-151	-240	-400
-146	-230	-382
-140	-220	-364
-134	-210	-346
-129	-200	-328
-123	-190	-310
-118	-180	-292
-112	-170	-274
-107	-160	-256
-101	-150	-238
-96	-140	-220
-90	-130	-202
-84	-120	-184
-79	-110	-166
-73	-100	-148
-68	-90	-130
-62	-80	-112
-57	-70	-94
-51	-60	-76
-46	-50	-58
-40	-40	-40
-34	-30	-22
-29	-20	-4
-23	-10	+14
-17.8	0	+32

°C		°F
-5.0	23	73.4
-4.4	24	75.2
-3.9	25	77.0
-3.3	26	78.8
-2.8	27	80.6
-2.2	28	82.4
-1.7	29	84.2
-1.1	30	86.0
-0.6	31	87.8
0.0	32	89.6
0.6	33	91.4
1.1	34	93.2
1.7	35	95.0
2.2	36	96.8
2.8	37	98.6
3.3	38	100.4
3.9	39	102.2
4.4	40	104.0
5.0	41	105.8
5.6	42	107.6
6.1	43	109.4
6.7	44	111.2
7.2	45	113.0
7.8	46	114.8
8.3	47	116.6
8.9	48	118.4
9.4	49	120.2

°C		°F
22.8	73	163.4
23.3	74	165.2
23.9	75	167.0
24.4	76	168.8
25.0	77	170.6
25.6	78	172.4
26.1	79	174.2
26.7	80	176.0
27.2	81	177.8
27.8	82	179.6
28.3	83	181.4
28.9	84	183.2
29.4	85	185.0
30.0	86	186.8
30.6	87	188.6
31.1	88	190.4
31.7	89	192.2
32.2	90	194.0
32.8	91	195.8
33.3	92	197.6
33.9	93	199.4
34.4	94	201.2
35.0	95	203.0
35.6	96	204.8
36.1	97	206.6
36.7	98	208.4
37.2	99	210.2
37.8	100	212.0

Table A.E.2

100 to 1000

C		F	C		F
38	100	212	260	500	932
43	110	230	266	510	950
49	120	248	271	520	968
54	130	266	277	530	986
60	140	284	282	540	1004
66	150	302	288	550	1022
71	160	320	293	560	1040
77	170	338	299	570	1058
82	180	356	304	580	1076
88	190	374	310	590	1094
93	200	392	316	600	1112
99	210	410	321	610	1130
100	212	413.6	327	620	1148
104	220	428	332	630	1166
110	230	446	338	640	1184
116	240	464	343	650	1202
121	250	482	349	660	1220
127	260	500	354	670	1238
132	270	518	360	680	1256
138	280	536	366	690	1274
143	290	554	371	700	1292
149	300	572	377	710	1310
154	310	590	382	720	1328
160	320	608	388	730	1346

1000 to 2000

C		F	C		F
538	1000	1832	816	1500	2732
543	1010	1850	821	1510	2750
549	1020	1868	827	1520	2768
554	1030	1886	832	1530	2786
560	1040	1904	838	1540	2804
566	1050	1922	843	1550	2822
571	1060	1940	849	1560	2840
577	1070	1958	854	1570	2858
582	1080	1976	860	1580	2876
588	1090	1994	866	1590	2894
593	1100	2012	871	1600	2912
599	1110	2030	877	1610	2930
604	1120	2048	882	1620	2948
610	1130	2066	888	1630	2966
616	1140	2084	893	1640	2984
621	1150	2102	899	1650	3002
627	1160	2120	904	1660	3020
632	1170	2138	910	1670	3038
638	1180	2156	916	1680	3056
643	1190	2174	921	1690	3074
649	1200	2192	927	1700	3092
654	1210	2210	932	1710	3110
660	1220	2228	938	1720	3128
666	1230	2246	943	1730	3146

°C		°F
166	330	626
171	340	644
177	350	662
182	360	680
188	370	698
193	380	716
199	390	734
204	400	752
210	410	770
216	420	788
221	430	806
227	440	824
232	450	842
238	460	860
243	470	878
249	480	896
254	490	914

°C		°F
393	740	1364
399	750	1382
404	760	1400
410	770	1418
416	780	1436
421	790	1454
427	800	1472
432	810	1490
438	820	1508
443	830	1526
449	840	1544
454	850	1562
460	860	1580
466	870	1598
471	880	1616
477	890	1634
482	900	1652
488	910	1670
493	920	1688
499	930	1706
504	940	1724
510	950	1742
516	960	1760
521	970	1778
527	980	1796
532	990	1814
538	1000	1832

°C		°F
671	1240	2264
677	1250	2282
682	1260	2300
688	1270	2318
693	1280	2336
699	1290	2354
704	1300	2372
710	1310	2390
716	1320	2408
721	1330	2426
727	1340	2444
732	1350	2462
738	1360	2480
743	1370	2498
749	1380	2516
754	1390	2534
760	1400	2552
766	1410	2570
771	1420	2588
777	1430	2606
782	1440	2624
788	1450	2642
793	1460	2660
799	1470	2678
804	1480	2696
810	1490	2714

°C		°F
949	1740	3164
954	1750	3182
960	1760	3200
966	1770	3218
971	1780	3236
977	1790	3254
982	1800	3272
988	1810	3290
993	1820	3308
999	1830	3326
1004	1840	3344
1010	1850	3362
1016	1860	3380
1021	1870	3398
1027	1880	3416
1032	1890	3434
1038	1900	3452
1043	1910	3470
1049	1920	3488
1054	1930	3506
1060	1940	3524
1066	1950	3542
1071	1960	3560
1077	1970	3578
1082	1980	3596
1088	1990	3614
1093	2000	3632

REFERENCES

Chapter 1

Indium Corporation of America, "Indalloy Specialty Solders," Form No. 102-8655 CP10M, Utica, N.Y., rev. May 1980.

Manko, Howard H., *Solders and Soldering*, McGraw-Hill, New York, 1964.

Chapter 2

Borase, Vijay, "Substrates Influence Thin Film Performance," *Microwaves*, October 1982.

———— "Electrical Properties Govern Substrate Effectiveness," *Microwave & RF*, February 1983.

Donegan, Tom, "PTFE Substrates Require Special Care in Fabrication," *Microwave*, October 1982.

Frey, Jeffrey, *Microwave Integrated Circuits*, Artech House, Dedham, Mass., 1975.

Fogiel, M., *Modern Microelectronic Circuit Design, IC Applications, Fabrication Technology*, vol. 1, Staff of Research and Education Association, New York, 1981.

Graff, Rudolf F., *Modern Dictionary of Electronics*, 5th ed., Howard W. Sams and Co., Indianapolis, 1982.

Materials Research Corp., *The Basics of Materials for Thin Films*, Orangeburg, N.Y.

―――― *The Basics of Sputtering*, 3rd ed. Orangeburg, N.Y.

Nowicki, Thomas E., "Microwave Substrates Present and Future," in *Microwave Tech Topics*, vol. 2, 3M Company, St. Paul, Minn..

Olyphant, Murray, "Measuring Anisotrophy in Microwave Systems," IEEE MTT-S, 1979.

Olyphant, Murray, and Thomas E. Nowicki, "Microwave Substrates Support MIC Technology," vols. 1 and 2, *Microwave*, November and December 1980.

Olyphant, Murray, D.D. Demeny, and Thomas E. Nowicki, "Epsilam 10 —A New High Dielectric Constant Conformable Copper-Clad Laminate," Cutips No. 6, 3M Company, St. Paul, Minn.

Rogers Corporation, "The Advantage of Nearly Isotropic Dielectric Constant for RT/Duroid 5870-80 Glass Microfiber–PTFE Composite," Chandler, Ariz., 1981.

Reference Data for Radio Engineers, 6th ed., ITT, Howard W. Sams, Indianapolis, 1975.

Vossberg, Walter, A., "Stripping the Mystery from Stripline Laminates," *Microwaves*, January 1968.

Chapter 3

Graff, Rudolf F., *Modern Dictionary of Electronics*, 5th ed., Howard W. Sams, Indianapolis, 1982.

Indium Corporation of America, *Understanding Solders and Soldering*, Utica, N.Y., 1981.

Laverghetta, Thomas S., *Microwave Measurements and Techniques*, Artech House, Dedham, Mass., 1976.

Manko, Howard H., *Solders and Soldering*, McGraw-Hill, New York, 1964.

Scott, Ewing C., and Frank A. Kanda, *The Nature of Atoms and Molecules*, Harper & Row, New York, 1962.

Harper, Charles A., *Handbook of Thick Film Hybrid Microelectronics*, McGraw-Hill, New York, 1974.

Isacoson, Nils I., "Automated Mask Generation: A Price/Performance and Practical Discussion," MSAT 83, Washington, D.C., March 1983.

State of the Art, Inc., *The Fundamentals of Thick Film Hybrid Technology*, ch. 16, 1970.

Chapter 5

Harper, Charles A., *Handbook of Thick Film Hybrid Microelectronics*, McGraw-Hill, New York, 1974.

MacDermid, Inc., "Metrex^R Etchant MU," Technical Data Sheet No. 9114.

Reference Data for Radio Engineers, 6th ed., Howard W. Sams, Indianapolis, 1975.

Standards and Specifications Committee of ISHM, "Glossary of Terms for Hybrid Microelectronic Standards Guidelines," ISHM STD02, Montgomery, Ala., April 1975.

Tramposch, Ralph, "Thin Film Processing of Hybrid IC's," *The Microwave Systems Designer's Handbook*, EW Communications, Inc., Palo Alto, Calif., 1983.

Vossberg, Walter, "Stripping the Mystery from Stripline Laminates," *Microwaves*, January 1968.

Chapter 6

Fogiel, M., *Modern Microelectronic Circuit Design, IC Applications, Fabrication Technology*, vol. 1, Staff of Research and Education Association, New York, 1981.

Graff, Rudolf., *Modern Dictionary of Electronics*, 5th ed., Howard Sams, Indianapolis, 1982.

Indium Corporation of America, *Understanding Solders and Soldering*, Utica, N.Y., 1981.

Johnson, D.R., and E.L. Chanez, *Characterization of the Thermosonic Wire Bonding Techniques*, Sandia Laboratories, Albuquerque, N.M.

Kester Solder Co., "Problems in Soldering Gold Plate," Laboratory Bulletin, January 1963.

Kulesza, Frank W., "New Epoxy Systems for Microelectronics," Epoxy Technology, Inc., Billerica, Mass., January 1976.

Manko, Howard H., *Solders and Soldering*, McGraw-Hill, New York, 1964.

Small Precision Tools, *Bonding Handbook*, Precision 79, San Rafael, Calif., 1977.

Sergent, Jerry, and E.C. Thompson, "Thick and Thin Film Hybrid Microcircuits Professional Advancement Course," Cahners Exposition Group, 1982.

3M Co., "A User's Guide to Vapor Phase Soldering with Flourimert^R Electronic Liquids," St. Paul, Minn., 1981.

Thwaites, C.J., *Soft Soldering Handbook*, International Tin Research Institute, Middlesex, England, 1982.

Yost, Frederick, G., "Soldering to Gold Films," *Gold Bulletin*, vol. 10, no. 4, October 1977.

Chapter 7

March, Steven, "Microstrip Packaging: Watch the Last Step," *Microwaves*, December 1981.

Markstein, Howard W., "Miniaturized Microwave Packaging — A Different Game," *Electronic Packaging and Production*, August 1982.

Chapter 8

MIL-C-39012C, "General Specification for Connectors, Coaxial, Radio Frequency," 11 August 1982.

MIL-C-39012C, Supplement 1, 30 September 1982.

Omni-Spectra, "Microwave Coaxial Connectors," Catalog, November 1980.

Wiltron Co., "46 GHz Coaxial Systems," Mountain View, Calif., January 1984.

INDEX

217

The Artech House Radar Library